ADHESIVES FOR THE
COMPOSITE WOOD PANEL INDUSTRY

ADHESIVES FOR THE COMPOSITE WOOD PANEL INDUSTRY

by

G.S. Koch, F. Klareich, B. Exstrum

Hagler, Bailly & Company
Washington, D.C.

NOYES DATA CORPORATION
Park Ridge, New Jersey, U.S.A.
1987

Copyright © 1987 by Noyes Data Corporation
Library of Congress Catalog Card Number 86-33300
ISBN: 0-8155-1115-9
Printed in the United States

Published in the United States of America by
Noyes Data Corporation
Mill Road, Park Ridge, New Jersey 07656

10 9 8 7 6 5 4 3 2 1

Library of Congress Cataloging-in-Publication Data

Koch, G.S.
    Adhesives for the composite wood panel industry.

    Bibliography: p.
    Includes index.
    1. Adhesives. 2. Wood --Bonding. I. Klareich, F.
II. Exstrum, B. III. Title.
TP968.K654 1987   674',83   86-33300
ISBN 0-8155-1115-9

Foreword

This book presents a market analysis of current fossil-fuel-based adhesives for the composite wood panel industry. It is also a study of the potential for less-energy-intensive biomass-derived adhesives for use in the industry. Adhesives manufacture and production account for a significant portion of overall wood panel industry energy use as well as overall production costs, and the wood panel industry consumes about 25% of the total U.S. adhesives production. Significant energy savings might be realized if current fossil fuel-based resins could be replaced with alternate biomass-derived adhesives. The book is an interesting study of potential alternative adhesives.

The majority of wood adhesives presently used in this industry are based on four major synthetic thermosetting resins: phenol-formaldehyde, resorcinol-formaldehyde, urea-formaldehyde, and melamine-formaldehyde. These four resins, which account for roughly 90% of all adhesive resins used in wood panels, are all derived from fossil fuels. The major nonconventional adhesives being investigated at this time as potential alternatives to thermosetting adhesives are: isocyanates, lignins, and tannins.

The book analyzes the current wood panel adhesives market, identifies domestic and international R&D efforts in the area of biomass-derived alternate adhesives which might substitute for conventional adhesives, and assesses technical and economic factors that will influence commercial success of these alternate adhesives.

The information in the book is from *Adhesives for the Composite Wood Panel Industry,* prepared by G.S. Koch, F. Klareich, and B. Exstrum of Hagler, Bailly & Company for the U.S. Department of Energy, January 1986.

v

The table of contents is organized in such a way as to serve as a subject index and provides easy access to the information contained in the book.

Advanced composition and production methods developed by Noyes Data Corporation are employed to bring this durably bound book to you in a minimum of time. Special techniques are used to close the gap between "manuscript" and "completed book." In order to keep the price of the book to a reasonable level, it has been partially reproduced by photo-offset directly from the original report and the cost saving passed on to the reader. Due to this method of publishing, certain portions of the book may be less legible than desired.

NOTICE

Contents and Subject Index

1 Introduction and Summary

BACKGROUND

The lumber and wood products industry, SIC 24, consumes
over 1.5 percent of the purchased fuels and electric
energy in the U.S. In the wood panel products industry, which consists mainly of plywood and particleboard
production, adhesives manufacture and application
account for a significant portion of overall wood panel
industry energy use as well as overall production
costs. The wood panel industry consumes about 25
percent of the total U.S. production of adhesives,
which translates into over 6 percent of all synthetic
binders consumed in the U.S. On average, plywood
consumes approximately 2 MMBtu/1000 square feet on a
3/8 inch basis, and particleboard consumes 2-10 times
this amount depending on the composition of the board.
(See Exhibit 1.a.)

Currently, the majority of wood panel ahesives are
based on four major synthetic thermosetting resins:

- phenol-formaldehyde
- resorcinol-formaldehyde
- urea-formaldehyde
- melamine-formaldehyde.

These four resins, which account for roughly 90 percent
of all adhesive resins used in wood panels, are all
derived from fossil fuels. Phenol and resorcinol are
derivatives of benzene which is synthesized from oil;
urea, melamine, and formaldehyde are all derivatives of
natural gas.

Significant energy savings could be realized if current
fossil fuel-based resins could be replaced with alternative biomass-derived adhesives. For this reason, the
U.S. Department of Energy's Office of Industrial Programs (OIP) engaged Hagler, Bailly & Company to analyze
the current wood panel adhesives market, identify both
domestic and international R&D efforts in the area of
biomass-derived alternative adhesives that might serve
as substitutes for conventional fossil fuel-based

1

Exhibit 1.a

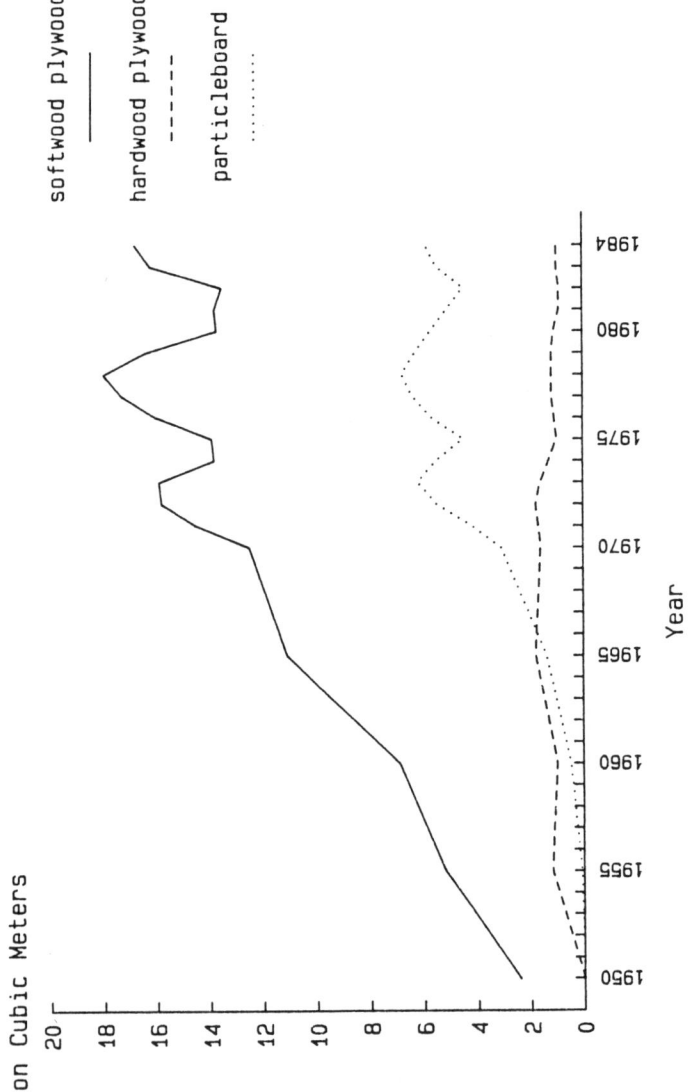

U.S. Wood-Based Panel Production

softwood plywood

hardwood plywood

particleboard

Million Cubic Meters

Year

Source: U.S. Department of Commerce

adhesives, and assess the technical and economic
factors that will influence commercial success of these
alternative adhesives. The results of this study will
be used to guide future program planning efforts for
the support of R&D in biomass-derived wood adhesives.

This market analysis is based upon information collect-
ed through both an extensive literature review and
interviews with key players in the wood panel industry.
These key players included universities, trade associations and professional societies, government agencies
and laboratories, as well as both users and suppliers
of wood adhesives in the private sector.

FINDINGS

Wood Panel Products

In 1984, the wood panel industry shipped approximately
$6.3 billion of plywood and particleboard, with plywood
accounting for roughly 90 percent and particleboard
accounting for 10 percent of this value. These dollar
values translate into a plywood production of 20.1
billion square feet on a 3/8 inch basis (17.82 million
cubic meters) and a particleboard production of approximately 3.2 billion square feet on a 3/4 inch basis
(5.664 million cubic meters).

The construction industry, where most of these wood
panel products are used, is forecast to grow at a
moderate rate of 3-4 percent. Within the wood panel
market itself, the market share held by individual
products will change. Plywood will continue to experience strong competition from cheaper imports as well as
substitution by composites, plastics, metals, and
cheaper wood products such as particleboard. Particleboard and other wood composite panels will enjoy
stronger growth in the market because of their lower
overall price.

Conventional Wood Adhesives

Currently, most of the adhesives used in the manufacture of these wood panel products are synthetic
thermosetting resins -- phenol-formaldehyde,
resorcinol-formaldehyde, urea-formaldehyde, and
melamine-formaldehyde -- meaning they require heat for
final setting. Although resorcinol and melamine resins

will cure at ambient temperatures, heat is usually applied to speed the curing process. The current U.S. market for thermosetting resins is shown in Exhibit 1.b.

Resorcinol-formaldehyde (RF) is known as the ideal wood adhesive because it forms the strongest wood bond, is waterproof, and is cold-setting. However, RF is the most expensive of the four resins, and is therefore limited in use to mixtures with phenol-formaldehyde resins or to specialty applications, such as marine construction, where a premium can be paid to obtain its waterproof properties. In 1983, roughly 3 million lb of RF resins were consumed by the wood panel industry.

Phenol-formaldehyde (PF) resins capture a large share of the wood panel adhesives market, mainly in the manufacture of plywood for exterior uses. These resins held the second-largest market share among the four thermosetting resins, with 594 million lbs (or 45 percent of total U.S. production) going into the wood panel products industry. PF resins are highly water resistant, but do require heat to cure. Roughly 45 percent of the U.S. sales of phenolic resins are for wood adhesives.

Urea-formaldehyde, used mainly in particleboard but also to a lesser extent in interior plywoods, holds the largest wood panel adhesive market share at 880 million lb, or 75 percent of total U.S. production in 1983. Although U.S. plywood production is six times larger than particleboard production, UF resin consumption is still larger than PF resin consumption because parti-cleboard on average consumes three times as much adhesive as plywood. Urea-formaldehyde (UF) is the cheapest of the four resins, but also yields the poorest adhesive bond, which can break down upon contact with water. UF resins have been heavily regulated in many European countries because of their tendency to outgas formaldehyde. This outgassing is caused by excess formaldehyde in the adhesive as well as by formaldehyde released when the UF bonds break down upon exposure to humidity. In the U.S., the Department of Housing and Urban Development and the Environmental Protection Agency are seriously consider-ing the initiation of similar regulations, particularly aimed at reducing UF resin use in mobile home construc-tion.

Exhibit 1.b

U.S. Production of Thermosetting Resins (1983)

Resin	Production (million lb, dry weight)		% Used in Wood Products
	Total	For Wood Panel Products	
Resorcinol-formalde-hyde	31	3	11
Urea-formaldehyde	1,172	880	75
Melamine-formaldehyde	180	11	6
Phenol-formaldehyde	1,320	594	45
Totals	2,703	1,488	55

Source: U.S. Department of Commerce

Melamine-formaldehyde (MF) is more water resistant than UF resins. However, it is relatively expensive and is therefore used mainly in combination with cheaper UF resins. Because of its white color, MF is also used extensively in specialty products where a light color is important. Although total MF resin production is high (180 million lb in 1983), only 6 percent or 11 million lb is consumed by the wood panel industry. The largest consumer of these MF resins is in molding compounds.

Taking into account the feedstock and process energies required to produce the two predominant resins, phenol-formaldehyde and urea-formaldehyde, the overall energy consumption in wood panel adhesives can be estimated. Using 1984 production figures, approximately .067 quad was consumed in the U.S. production of phenolic resins. Roughly 65 percent of these PF resins were used as binders in wood products, yielding a total PF energy consumption of .043 quad for wood adhesives. In 1984, approximately .021 quad was consumed in the production of UF resins, and with 75 percent of these resins consumed in wood adhesives, a total of .016 quad can be estimated as the energy (feedstock and process) consumed by UF resins in wood adhesives.

Potential Alternative Adhesives

The major non-conventional adhesives being investigated at this time as potential alternatives to thermosetting adhesives are:

- isocyanates
- lignins
- tannins.

Isocyanates are petroleum-derived products, and therefore result in virtually no net reduction in the embodied energy in wood adhesives. However, isocyanates are closest to becoming commercially available in the U.S., with significant market penetration expected in the next 5 years. One problem remaining with isocyanates is their characteristic tackiness, which can result in their sticking to the press cauls. Industry representatives believe, however, that recently developed release agents will solve this problem.

Lignins are currently being used as wood adhesive
extenders. However, research is focusing on two major
routes for substituting lignins for current petroleum-
based adhesive resins:

- direct substitution of lignins for phenolics

- breakdown of lignins into phenolic precursors
  which can then be synthesized into phenolics.

Although much research remains in both of these areas,
there is a fundamental limitation in the available
reactive sites on the lignin molecule that will limit
the bonding quality of directly-substituted lignin
adhesives, thus reducing their competitiveness with
conventional adhesives. In addition, the lignin
molecules that can be extracted from wood residues vary
greatly depending on chemical processing. This results
in varying adhesive compositions that lead to varying
adhesive bond qualities, an unacceptable condition for
panel producers. Although 100 percent lignin substitution suffers from these disadvantages, industry
representatives believe that partial substitution at
roughly 40 percent will be more feasible. One major
advantage of direct lignin utilization is that the wood
industry would no longer be dependent on outside
sources for its adhesive materials, since it would be
using residues from a wood-related process.

When lignins are broken down into phenolic precursors
and built back up into phenol, a more consistent, more
reactive adhesive can be made. Such an adhesive would
be very similar to a conventional thermosetting adhesive. This type of process, however, would require the
addition of new chemical processing equipment, which
may be a hinderance to acceptance by the wood industry.

Tannins, usually derived from bark or other wood
residues, could also replace the petroleum-based
thermosetting adhesives in wood panel applications. In
the case of tannins, research is needed to control or
slow down the highly-reactive tannin molecules. Commercial tannin facilities based on wattle do exist today
in South Africa. However, this experience cannot be
directly applied in the U.S. because the quality of the
tannin adhesive, like lignin adhesives, changes as the
biomass raw materials change. Tannin production
requires large, centralized production facilities, and

even then, it is uncertain that the economics of such a
facility will prove to be feasible.

Both lignin and tannin adhesives would lower feedstock
fuel use by replacing conventional thermosetting
resins. However, both types of adhesives require
additional process energy for extraction or modification. In addition, these biomass products are
currently burned for energy or used elsewhere in the
paper and wood products industry. If they are used for
adhesives, their energy contribution will have to be
replaced with some other fuel. Although coal has been
suggested as a potential replacement, industry representatives believe that oil is the more likely candidate
because it is easier to handle.

Exhibit 1.c summarizes the R&D needs and energy implications for the three major alternative adhesives.

Criteria for Success

Many trade-offs exist between isocyanate, lignin, and
tannin adhesives. Even if the technical problems
outlined above are solved, several key criteria will
have to be satisfied before these alternative adhesives
can become commercially feasible. In order of relative
importance, the major wood panel industry acceptance
criteria are as follows:

● Consistency -- as relates to the reliability
of the bond. Usually requires a chemically
clean product so that an adhesive mix will
have predictable performance characteristics.

● Cost -- a factor since the industry must keep
down the price of its panel products by using
inexpensive adhesives. There are only a
limited number of end-use applications where
the consumer will pay a premium for better
performance. However, the industry will not
accept a lower-quality adhesive even if the
price is somewhat lower than conventional
adhesives. Therefore, cost is not the
overriding criteria for acceptance.

● Level of panel production -- the industry
will not tolerate longer cure or press times
than those experienced with conventional

Exhibit 1.c

Major R&D Needs and Potential Energy Benefits (Losses)

Alternative Adhesive	Major R&D Need	Potential Energy Conservation Gains/Losses		
		Feedstock	Process	Total Bond Cost
Isocyanates	• Development of release agents	No change	Loss - complex multistage processing	Gain - faster setting resin
Lignins:				
- Direct substitution	• Increase reactivity; • Generate a uniform supply	Potential gain, although may need a replacement fuel source	Loss - purification and modification may be energy intensive	Loss - longer press times
- Breakdown to phenolic precursors	• Improve process efficiency to enhance economics	Potential gain, although may need a replacement fuel source	No change	No change
Tannins	• Slow down reactivity; • Generate a uniform product	Gain	Loss - extraction may be energy intensive	Gain - faster set resin

NOTE: Feedstock - refers to the embodied energies of the raw materials

Process - refers to the energy used in formulating the resin or isolating adhesive materials.

Total Bond Cost - refers to the energy costs associated with bonding, i.e., cure and press times, cure temperatures. These are the only energy costs directly born by the wood panel producer.

Source: Hagler, Bailly & Company, through data compiled from extensive industry interviews.

adhesives. Such extended cure or press times
could significantly lower panel production,
and thus lower the economic return.

- **Equipment changes** -- equipment changes are
  costly to the wood panel industry, and since
  it is a depressed industry, few if any
  equipment changes will be tolerated.

These factors combine to yield a **total bond cost**, which
is the cost that will determine the commercial success
or failure of an alternative adhesive.

It is likely that some or all of the alternative
adhesives will enter the market in the next 5-10 years.
Much fundamental research must first be carried out to
characterize domestic biomass sources and develop an
understanding of how to inexpensively control their
reactivity and form a dependable adhesive bond. However, once these technical and economic problems have
been solved, biomass-derived adhesives for the wood
panel industry could play an important role in
conserving scarce U.S. fossil fuel resources by
substituting renewable biomass for current fossil fuel-
based raw materials. Based on 1984 production rates,
if 100 percent of the UF and PF resins were replaced, a
potential savings of roughly .059 quad (minus the energy consumed in processing the alternative adhesives)
could be realized. In this study, it was not possible
to quantify the exact energy-saving potential of all
the various alternative adhesive systems. This remains
to be done on a case-by-case basis.

ORGANIZATION OF REPORT

Chapter 2 presents a general overview of the U.S. wood
panel industry. Chapter 3 describes the conventional
wood adhesives currently used in this industry. R&D
for alternative non-conventional wood adhesives is
outlined in Chapter 4, along with a summary description
of key R&D players. Chapter 5 provides an outlook on
the future of the wood panel industry, potential
changes in adhesive use, and the technical and economic
potential of the alternative non-conventional adhesives.

Appendix A provides detailed market data. References
are listed in Appendix B, and a glossary is provided in
Appendix C.

Appendix A provides more detailed market tables on
conventional adhesives. Appendix B lists the references used in this study. (Note that reference numbers
cited in the text refer to the numbered references in
Appendix B.) Appendix C contains a glossary of key
terms used in this report.

2 The Composite Wood Panel Industry

The U.S. lumber and wood products industry, SIC 24, purchased 184.8 trillion Btu of fuels and electric energy in 1981 at a cost of $1,163 million. This translates into roughly 1.6 percent of the total industrial energy purchases for heat and power.[139] These energy consumption figures, however, do not incorporate feedstock energies, for example the embodied energy of adhesives and binders. From 1978–1980, the wood products industry annually consumed 2.3 billion lb of adhesives, or roughly 25 percent of all the adhesives and over 6 percent of all synthetic binders consumed in the U.S.[119]

Shipments of all industries within SIC 24 are presented in Exhibit 2.a, which shows 1984 total shipments of almost $19.7 billion (in 1972 dollars). As shown in Exhibit 2.a, the lumber and wood products industry can be divided into several subgroups by product type, including logging, lumber, hardwood flooring, structural wood members, millwork, pallets and skids, wood preservation, plywood, and particleboard. The wood panel products industry consists of three of these subgroups:

- SIC 2435   hardwood plywood

- SIC 2436   softwood plywood

- SIC 2492   particleboard (including waferboard, strand board, chipboard, fiberboard, flakeboard, and other reconstituted wood panels)

The combined shipments of these three subgroups is estimated at $6.3 billion ($2.6 billion in 1972 dollars) for 1984 (see Exhibit 2.b). Wood panel shipments (expressed in constant 1972 dollars) rose by 9.8 percent from 1983 to 1984, with a breakdown among individual panel types as follows:

Exhibit 2.a

Wood Products Industry Shipments: 1982 - 1985
(values in million 1972 dollars)

SIC	Industry	1982	1983	1984	Percent Change 83-84	1985	Percent Change 84-85
2411	Logging operators	4.258 9	4.439 9	4.600 0	3 6%	4.462 0	-3 0%
2421	Lumber	5.365 0	5.500 2	6.190 8	12 6%	6.098 0	-1 5%
2426	Hardwood dimen flooring	418 2	460 2	482 8	4 9%	463 9	-4 0%
2439	Structural wood members	459 6	446 5	471 2	5 5%	466 5	-0 1%
2431	Millwork	1.786 7	2.542 5	2.623 9	3 2%	2.571 0	-2 0%
2448	Pallets & Skids	656 9	741 9	816 0	9 9%	848 0	3 9%
2491	Wood preservation	641 9	739 9	769 8	4 0%	770 0	0.03%
2435	Hardwood veneer & plywood	753 5	840 0	850 0	1 2%	830 0	-2 4%
2436	Softwood veneer & plywood	1.823 0	2.239 2	2.592 0	15 8%	2.400 0	-7 4%
2492	Particleboard	259 5	276 4	295 7	7 0%	305 0	3 1%

Source: U.S. Department of Commerce, U.S. Industrial Outlook 1985.

Exhibit 2.b

Wood Panel Product Trends

Item	1972	1977	1979	1981	1982	1984'	Compound annual rate of growth 1972-84	1985'	Percent change 1984-85
Industry Data									
Value of Shipments²	3.218 7	5.508 1	6.389 4	5.754 7	5.070	7.072	6 8	—	—
Softwood veneer & plywood (SIC 2436)	2,011.5	3.804.8	4.295.3	3.887.8	3.207	4.900	7.7	—	—
Hardwood veneer & plywood (SIC 2435)	911 8	1.250 4	1.554.3	1.534 7	1.326	1.800	4.8	—	—
Particleboard (SIC 2492)	295 4	452 9	539 8	543.2	537.4	859 3	6.9	—	—
Value of Shipments (1972$)	3.218 7	3.256 2	3.238 4	3.013 8	2.836	3.736	1.2	3.535	−5 4
Softwood veneer and plywood	2.011.5	2.069 0	2.038 6	1.907 8	1.823	2.593	2.1	2.400	−7 4
Hardwood veneer and plywood	911 8	868 3	869.8	846.5	753 5	899	−0.2	990	0 1
Particleboard	295 4	318 9	330 0	259 5	259.5	293	−0 1	305	4 1
Total employment (000)	76 5	74 7	77 8	77.2	58 4	65 4	−1.3	66 4	5 6
Softwood veneer and plywood	43 7	46 2	47 6	39 4	34 8	37 7	−1.3	37 0	1 9
Hardwood veneer and plywood	25 1	22 3	24 2	22 7	18 0	21.3	−1.4	21 0	1 4
Particleboard	7 7	6 2	6 0	5 1	5.5	6 4	−1.5	6 6	3 1
Production workers (000)	68 4	66 0	69 6	59.3	50 7	58 4	−1.3	58 9	5 9
Softwood veneer and plywood	39 9	41 9	43 3	35.3	31.0	34.9	−1.1	34 2	−2 0
Hardwood veneer and plywood	21 1	19 1	21.2	19 6	15.3	18 4	−1.3	18 2	−1 1
Particleboard	6 4	5 0	5 1	4 4	4.4	5.2	−1 7	5 4	3 8
Average hourly earning of production workers	3 69	5 61	6 77	7 76	8 15	8 80	7.5	—	—
Softwood veneer and plywood	4 18	6 23	7 74	8 83	8 93	9 74	7.3	—	—
Hardwood veneer and plywood	2 80	4.27	4 88	5 77	6.24	8 79	7 7	—	—
Particleboard	3 75	5 53	6 41	8 00	8 67	9 44	8 3	—	—
Capital expenditures	132 1	167 7	202 2	242 9	156 4	—	...	—	—
Softwood veneer and plywood	65 4	105 6	144 0	187 1	99 7	—	...	—	—
Hardwood veneer and plywood	32 1	29 6	32 5	33 4	20 7	—	–	—	—
Particleboard	34 6	32.5	25.7	22 4	36 0	—	–	—	—
Product Data									
Value of shipments⁴	3.107 7	5.218 8	6.040 1	5.133 1	4.573	6.275	6.0	—	—
Softwood veneer and plywood	1.939 6	3.582 3	4.045 0	3.201 2	2.767	4.260	6.8	—	—
Hardwood veneer and plywood	875 3	1.160 0	1.426 9	1.373 7	1.262	1.425	4 1	—	—
Particleboard	292 8	476 5	568 4	558.2	544 8	668 7	7.1	—	—
Value of Shipments (1972$)	3.107 7	3.089 2	3.065 7	2.683 4	2.553	3.315	0 5	3.134	−5 5
Softwood veneer and plywood	1.939 6	1.948 0	1.919 8	1.856 1	1.573	2.254	1.3	2.086	−7 5
Hardwood veneer & plywood	875 3	805 6	798 5	757.7	716.8	801	−0 7	800	−0 1
Particleboard	292 8	335 6	347.4	272.2	263 1	297.2	0 1	309 1	4 0
Product price index (1972 = 100)	—	—	—	—	—	—	—	—	—
Softwood veneer and plywood	100 0	187.5	211.5	197 3	179.0	203 5	6 1	—	—
Hardwood veneer and plywood	100 0	145 6	179 2	181 4	175 9	189 4	9 5	—	—
Particleboard	100 0	142 0	163.6	205 1	207.1	230	7.2	—	—
Quantity shipped	—	—	—	—	—	—	—	—	—
Softwood veneer and plywood (mil sq ft 3/8)	17.974 7	19.524 0	18.499 8	15.529 4	14.164 5	—	—	18.306	5 0
Hardwood veneer and plywood (mil sq ft 3/8)	4.950 4	3.939 4	3.550.1	3.300 0	2.915 7	—	—	3.687	4 6
Particleboard (mil sq ft 3/4) (excludes structural particleboard and medium density fiberboard)	3.015	3.554	3.376	2.653	2.343	—	—	2.974	4 9
Trade Data									
Value of exports	67 7	139 0	219 7	316 0	224 4	280 5	12.6	284	1 4
Softwood veneer and plywood	46 2	80 1	11 8	184 6	118 6	138 2	9 6	130 0	−5 9
Hardwood veneer and plywood	15 2	46 3	88 3	96 5	86 2	112.0	18 1	120	7 1
Particleboard	6 3	12.6	19.6	34 9	19 6	30 3	14.0	34 0	12 2
Value of imports	408 8	575 2	727 9	830 0	474 2	890 1	4.5	684	− 8
Softwood veneer and plywood	8 4	25 7	39 8	35 9	29.6	37 7	13.3	36	4 5
Hardwood veneer and plywood	398 5	512 6	633 0	532 9	391 3	538 7	2 5	525	−2 5
Particleboard	1.9	36 9	55 1	61 2	53.3	113 7	40 6	118 0	3 8
Exports/shipments ratio	0.022	0 027	0 036	0.061	0.049	0.043	5.7	0.047	3 3
Softwood veneer and plywood	0.023	0 020	0 025	0.051	0.043	0.032	2.8	0.033	1 7
Hardwood veneer and plywood	0.017	0.066	0 096	0.070	0.068	0.079	13.7	0 084	7 1
Particleboard	0.022	0.026	0 034	0.063	0.036	0.045	8 1	0 049	8 0
Import/new supply ratio	0.116	0.099	0 108	0 109	0.094	0.098	−1.4	0 10	2 0
Softwood veneer and plywood	0.004	0.007	0 010	0 011	0.009	0.008	5 9	0 008	3 6
Hardwood veneer and plywood	0.321	0.321	0.333	0.300	0 264	0.296	−0 7	0 292	−1 3
Particleboard	0 006	0.072	0 088	0.099	0 089	0 145	30 4	0 145	−0 2

' Estimated
² Forecast
³ Value of all products and services sold by SIC 2435, 2436 and 2492
⁴ Value of shipments of softwood veneer, hardwood plywood and particleboard products produced by all industries

⁵ New supply is the sum of product shipments plus imports
Source Bureau of the Census and Trade Development/ITA Estimates and forecasts by the
Bureau of Industrial Economics

Source: U.S. Department of Commerce, <u>U.S. Industrial Outlook 1985</u>.

Panel Type	% of Total 1984 Shipments	Real Growth (1983-1984)
hardwood plywood	22	7.0 %
softwood plywood	68	11.3 %
particleboard	9	6.0 %

In general, the proportion of total shipments accounted for by plywood has declined over the last decade, mainly because of the growing use of reconstituted or composite wood products. By reconstituting small units of wood -- fibers, strands, particles, flakes, or wafers -- with binders or adhesives, composite wood products can use a larger portion of the tree resource and can also use a poorer quality wood than most solid wood products. The declining size of available logs (due to the use of younger trees) and the economic need to recover higher percentages of usable material from forest resources have spurred the growth of these bonded composite wood products. The manufacture of wood composites has also allowed the development of wood products with greater uniformity, reliability, and versatility than many solid wood products. The continuing impact of these factors means that the bonding of wood will become even more dominant in the wood processing industry in future years.[141]

In the following sections, the two major wood panel products -- plywood and particleboard -- are characterized in more detail, with an emphasis on adhesive use. This is followed by a discussion of the future of the U.S. wood panel industry and the major factors that will influence future markets.

PLYWOOD

Plywood is an assembly of layers of wood (veneers or plies) joined together by means of an adhesive. Generally plywood is formed by gluing together an odd number of plies with the grain of one ply at right angles to the next, thereby increasing the dimensional stability of the end-product. In some instances, a core plywood is made by gluing thin sheets of veneer onto a central thick base layer of wood, resulting in panels such as blockboard, laminboard, battenboard, and veneered wood.

Plywood is categorized by the type of tree from which the wood veneers are derived, either softwood or hardwood. Hardwood plywood is comprised of wood from a broad-leaf tree such as oak, walnut, or maple, and is generally used in decorative building applications such as exterior siding, interior wall panels, furniture, laminated block flooring, and flush doors. It also has some limited uses in boat construction, aircraft parts, and sporting goods. Most hardwood plywood panels range in thickness between 1/4 and 1/2 inch. The standard U.S. measure for plywood production, however, is surface area with an assumed standard thickness of 3/8 inch.

The veneers for softwood plywood come from coniferous trees and is mainly employed in structural applications. It ranges in thickness from 1/4 to 1-1/4 inches, with a standard thickness of 3/8 inch.

Manufacture

The manufacture of hardwood and softwood plywood is fundamentally the same. For hardwood plywood, the wood veneers are dried and the adhesive is applied by rubber-covered rollers that are inside glue spreaders. Alternate wood veneers are passed through the glue spreader, which coats both sides of the veneer. The coated veneers are then interlayered with uncoated veneers for final assembly in preparation for curing. The assembled veneers are pressed in a multiple entry hot-press at approximately 250°F and 150-300 psi. Press times are roughly 5-7 minutes. Some adhesives cure at ambient temperature, but this procedure is normally not used in the United States (see Exhibit 2.c for process flowchart).

With softwood plywood manufacturing, the difference in absorptivity among the various wood species results in different levels of wet glue line drying before the panel can be pressed. Softwood veneer is normally dried to a very low moisture content. In a typical five-ply panel production process, the two cross-plies are run through high speed double roll spreaders. The film thickness on the rolls is controlled by a doctor roll or knife attached to a special glue hopper on the side of the roll. The veneers then undergo the same process described above for hardwood plywood, with double coating and assembly of alternate pieces of coated and uncoated veneers.

Exhibit 2.c

Manufacture of Hardwood Plywood

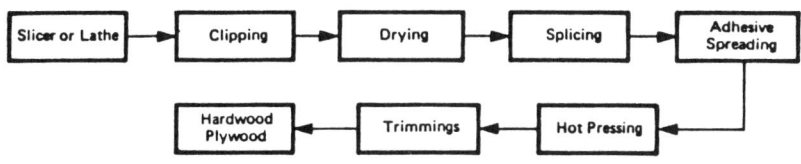

Source: Skeist, Irving Handbook of Adhesives, 2nd Edition, 1977.

The assembled plywood panels are next inserted into the various openings of a multiple entry hot-press, using special loading and unloading platforms. The panels are automatically fed into and out of the hot-press. Press cycles are dependent upon panel thickness, and are determined by the distance of the innermost glue line from the hot platen. Since wood transfers heat slowly, the greater the distance to the innermost glueline, the longer the press time.[119,125]

Adhesives Use

In exterior applications, plywood must meet weatherability standards. This requires waterproof adhesives such as phenol-formaldehyde (PF), that can withstand exposure not only to weather, but also to microorganisms, hot and cold water, and dry heat. These adhesives have proven to be more durable than the wood itself. In 1980, over 99 percent of U.S. structural softwood plywood production was bonded with phenol-formaldehyde adhesives, with a total PF adhesive consumption of approximately 500 million lb on a dry basis. A typical softwood plywood adhesive mix (on a percentage of total mix basis) is as follows:[119]

- PF resin solids   28.0 %
- filler   8.0 %
- extenders (wheat flour)   5.0 %
- caustic soda (NaOH), dry   1.5 %
- water   57.5 %

In contrast, decorative hardwood plywoods do not require the waterproof characteristics of PF adhesives. In 1980, over 95 percent of the U.S. decorative hardwood plywood production was bonded with urea-formaldehyde (UF) adhesives, with an estimated consumption of 60 million lb on a dry basis. A typical interior plywood adhesive mix would include (on a percentage of total mix basis):[119]

- UF resin solids   28.1 %
- extender (wheat flour)   21.6 %
- filler (nutshell flour)   5.4 %
- catalyst (solids)   0.6 %
- water   44.3 %

Interior grade and furniture plywood, both softwood and hardwood, are typically bonded with UF adhesives. One exception is southern softwoods, which must be bonded

with phenolic resins. These trees grow sporatically
during the year, resulting in springwood which is so
porous that it absorbs and rapidly dries out the
adhesive during pressing, and summerwood which has so
much pitch that the adhesive cannot effectively
penetrate the wood. Therefore, southern plywoods use
thick glue lines of phenolic resins. Since southern
plywood is growing more rapidly than other plywood
markets, there is a general industry trend toward
increasing in the overall adhesive use per square foot
of plywood. In 1984, overall plywood mill glue costs
typically averaged $11 per thousand square feet (MSF)
on a 3/8 inch basis.[25] (See Chapter 3 for more
information on PF and UF adhesives.)

Market

The bonding of softwood plywood represents one of the
largest single applications of adhesives in industry,
with total production of softwood plywood reaching
18,907,107 thousand square feet (16.73 million cubic
meters) on a 3/8 inch basis in 1984. Hardwood plywood
is a depressed industry in the United States -- it is
encountering increasing amounts of foreign competition
within its own domestic market. As a result,
production of hardwood plywood reached only 1,226,889
thousand square feet, 3/8 inch basis in 1984 (1.09
million cubic meters).[140]

The softwood plywood industry is mainly concentrated in
the Pacific Northwest (Washington, Oregon, and
California), where the major tree source is the Douglas
Fir, and the South (Louisiana, Alabama, and North
Carolina), where the major tree source is the Southern
Yellow Pine. The South is expected to be the major
growth area for softwood plywood production in the
future, although this growth will slow as the timber
base becomes more fully utilized.

The softwood plywood industry is not heavily
concentrated, with over 170 plants employing almost
38,000 employees. The top eight plywood producers
account for less than 60 percent of total U.S.
capacity. The largest producer is Georgia Pacific,
followed by Champion International, Boise-Cascade,
Williamette Industries, Weyerhaeuser, International
Paper, Southwest Forest Industries, and Louisiana-
Pacific.

The softwood plywood industry is relatively self-
sufficient, with imports capturing less than 1 percent
(major source is Canada) of the U.S. market. In 1984,
the end markets for softwood plywood were:

- new housing   40 %
- homeowner repair and remodeling   29 %
- nonresidential construction   14 %
- industrial   14 %
- exports   3 %.

This market distribution is expected to remain somewhat
stable over the coming years. Despite increasing
competition from particleboard, softwood plywood's
consistent high strength and moisture resistance will
result in a stable market with an expected 1.5 percent
annual growth rate. By 1995, the construction industry
is expected to consume over 18.2 billion square feet of
softwood plywood, with manufacturing accounting for an
additional 5.1 billion square feet of consumption.

The hardwood plywood industry is mainly concentrated in
the South (Georgia and Virginia), which accounts for
over 55 percent of U.S. production, and the Pacific
Northwest (Oregon), which accounts for approximately 25
percent of domestic output. The most common tree
species used include oak, birch, and gum, although the
lower-quality maple, mahogany, and walnut species are
growing rapidly in market share.[130]

The hardwood plywood industry consists of approximately
150 plants which employ over 18,000 people. The
individual producers tend to be the same as those in
the softwood plywood industry, with Georgia-Pacific
capturing the largest market share.

Hardwood plywood production is generally declining as
costs rise and domestic hardwoods become increasingly
scarce. Currently, imports capture over 75 percent of
the U.S. market. The major sources for these imports
are the Philippines, Korea, Indonesia, Malaysia, and
Taiwan.[130, 141]

PARTICLEBOARD

Particleboard is essentially composition board made up
of individual dry wood chips, flakes, splinters, or
strands. Included under the term particleboard is a

broad range of products that vary according to wood species, particle size and shape, and binder type. Depending on composition, particleboards are commonly referred to as either fiberboard, chipboard, flake board, and wood waste or sawdust boards. To produce particleboard, wood fragments are coated with a resin binder and formed into a particular shape by the application of heat and pressure. Many types of binders can be used in the formation of particleboard, including inorganic elements such as gypsum or thermosetting resins based on urea, melamine, or phenol. Particleboard can be formed in a single, homogeneous layer of wood particles or it can be multilayered.

Manufacture

Three basic processes are used to manufacture particleboard:

- single or multi-platen hydraulic hot-plate press

- extrusion between heated plates

- continuous pressing.

Exhibit 2.d presents a flow diagram of the manufacture of particleboard. The production of hydraulic hot-pressed particleboard as described below accounts for more than 90% of total output in the United States.

Urea resins are extensively used in making particleboard for interior applications where it is important that the binder be colorless and inexpensive. Typically, the production of particleboard with urea or melamine resins requires that the wood particles be conditioned to reduce their water content prior to spraying with the resin solution. The mixture of wood and binder is then spread in controlled amounts onto a uniform mat that rests on cauls or trays which move by conveyor. The mat is passed through a prepressing unit and onto a preloading rack. From the preloading rack, the mat goes into a multi-platen press. Mats are then pressed individually between heated platens in a hydraulic press to cure the binder. Press temperatures generally range between 250 and 350°F. The boards are removed from the press without cooling.

Exhibit 2.d
Manufacture of Particleboard

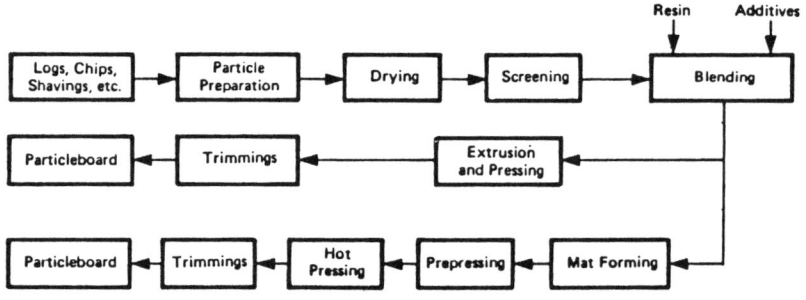

Source: Skeist, Irving Handbook of Adhesives,
          2nd Edition, 1977.

Upon removal from the press, the boards have achieved
sufficient flexural strength to allow for handling and
trimming. Full cure as measured by flexural strength
is not reached until after an aging and conditioning
period of about three days. Flexural strength will
vary with resin type, particle size, and moisture
content. Increased resin and moisture content yield
greater flexural strength.[85, 125]

Adhesives Use

Usually 6-10% of the particleboard end-product is
comprised of the weight of the adhesive (on a dry
basis). Typical glue costs in 1984 ranged from
$8.31/MSF for waferboard mills to $24.36/MSF for
oriented strandboard mills (on a 3/8 inch board basis).
The type of adhesive resin used in particleboard
manufacturing is largely a function of where the panel
will be used. Interior particleboard, which accounts
for over 90 percent of total particleboard production,
is usually not exposed to high moisture levels and can
therefore be bonded with inexpensive urea-formaldehyde
(UF) resins. Exterior particleboards require a higher
level of water resistance and are usually bonded with
phenolic resins.[25, 85]

Particleboards bonded with UF resins have a problem
with formaldehyde release that is not generally
associated with other resins. This is mainly due to
the release of excess formaldehyde and the breakdown of
urea-formaldehyde bonds upon exposure to heat and
moisture. This formaldehyde release becomes more
noticeable because these particular panels are used for
interior uses, so human exposure becomes more apparent.
In addition, since particleboards require more binder
than plywood for a given board volume, the formaldehyde
release from particleboard is apparently much higher
than from plywood in similar applications. To
partially control such formaldehyde release, melamine-
formaldehyde resins can be incorporated into the
adhesive mix.

In 1984, the U.S. Department of Housing and Urban
Development (HUD) developed standards for formaldehyde
emissions from hardwood plywood and particleboard. The
U.S. Environmental Protection Agency recently announced
standards for formaldehyde emissions in the workplace.
Most industry representatives feel that stricter
standards will be forthcoming in the next 5-7 years,

with the possibility that UF will be virtually eliminated from the interior wood panel market.[141]

Market

Particleboard has rapidly gained market share during the last decade with 1984 production reaching 3.2 billion square feet on 3/4 inch basis. Since particleboard is strong, resists warping and marring, and is easy to assemble without noticeable joints, it has quickly penetrated the flooring and furniture corestock markets. However, the value of chips and shavings as an energy source has also been increasing rapidly, and some industry representatives believe that the traditional price advantages of particleboard will disappear unless costs (including adhesive costs) are lowered.

Like plywood, the construction industry is the largest consumer of particleboard, accounting for roughly 55 percent of domestic consumption. The largest application in the construction industry is for floor underlayment, particularly in small homes and mobile homes. Of the manufacturing uses for particleboard, furniture corestock is the single largest application, accounting for over 70 percent of manufacturing consumption and almost 40 percent of total particleboard consumption.

The U.S. is a net importer of particleboard, with imports accounting for over 17 percent of total domestic consumption in 1984. Canada and Mexico were the major sources for these imports. Overall U.S. particleboard production is expected to grow by approximately 15 percent in real terms during the 1985-1989 time period.[141]

Approximately 55 U.S. plants manufacture particleboard, employing about 6,400 workers in 1984. Most of these plants are found in the Pacific Northwest and the South, close to timber supplies. Georgia-Pacific is the largest particleboard producer, followed by Weyerhaeuser and Williamette Industries. Other major players include Champion International, Louisiana-Pacific, Temple Eastex, Boise-Cascade, MacMillan-Bloedel Particleboard, Olinkraft, and Timber Products.[130]

FUTURE OF THE WOOD PANEL INDUSTRY

The wood panel industry is so closely tied with the
construction industry that their futures cannot be
separated. The U.S. Department of Agriculture's
forecasts for residential and nonresidential
construction through the year 2030 are shown in
Exhibits 2.e and 2.f. According to these projections,
the demand for plywood will nearly triple during this
time period, while the demand for particleboard will
rise by a factor of almost 4.5. The basic assumptions
used in these forecasts are shown in Exhibit 2.g.

In the near term, construction is expected to grow at a
modest 3-4 percent annual rate. Therefore, the overall
wood panel market will track this modest growth. Among
the individual panel products, hardwood plywood will
continue to decline at a slow pace as imports gain
market share and U.S. companies move out of the market
and into more profitable areas. The softwood plywood
market will continue to grow at a moderate 4 percent
annual rate in the near term, largely due to housing
starts and residential repair and remodeling. The
particleboard market is expected to grow at a rapid
rate, obtaining a 15 percent real market increase by
1989. Stricter environmental standards, however, will
probably force modifications in current particleboard
binders to reduce formaldehyde emissions.[141]

Exhibit 2.e

Use of Timber Products in New Housing: 1962 – 2030

Year	Lumber	Plywood (⅜-inch basis)	Board[a] (½-inch basis)
	Million board feet	*Million square feet*	*Million square feet*
1962.....	13,940	4,180	1,660
1970.....	12,270	6,330	2,070
1976.....	16,240	8,285	2,630
Low projections			
1990.....	21,845	11,060	4,900
2000.....	18,725	9,325	4,660
2010.....	19,800	9,855	5,120
2020.....	19,580	9,705	5,260
2030.....	15,645	7,755	4,270
Medium projections			
1990.....	23,345	11,825	5,200
2000.....	21,320	10,645	5,255
2010.....	22,540	11,235	5,790
2020.....	20,670	10,255	5,535
2030.....	17,910	8,900	4,855
High projections			
1990.....	24,715	12,520	5,500
2000.....	23,670	11,840	5,815
2010.....	26,580	13,270	6,805
2020.....	22,610	11,225	6,050
2030.....	23,065	11,495	6,210

[1]Includes both softwoods and hardwoods. Includes allowance for on-site and manufacturing waste.
[2]Hardboard, insulating board, and particleboard (including waferboard, flakeboard, composite board, and medium-density fiberboard).

Source: Forest Service, U.S. Department of Agriculture, <u>An Analysis of the Timber Situation in the United States 1952 – 2030</u>.

Exhibit 2.f

Timber Products Used in Nonresidential Construction: 1962 – 2030

Year	Lumber		Plywood (⅜-inch basis)		Board[1] (½-inch basis)	
	Total	Use per 1,000 dollars of expen-diture[a]	Total	Use per 1,000 dollars of expen-diture[a]	Total	Use per 1,000 dollars of expen-diture[a]
	Million board feet	*Board feet*	*Million square feet*	*Square feet*	*Million square feet*	*Square feet*
1962..........	3,303.4	57.2	1,639.4	28.4	605.0	10.5
1970..........	3,528.4	48.5	1,889.3	26.0	785.0	10.8
1973..........	3,695.3	49.8	2,158.5	29.1	915.3	12.3
1976..........	3,000.6	49.7	1,824.5	30.3	821.5	13.6
Low projections						
1990..........	4,090	42.7	3,080	32.1	1,610	16.8
2000..........	4,420	40.7	3,610	33.3	2,060	19.0
2010..........	4,750	39.2	4,160	34.3	2,510	20.6
2020..........	5,060	38.4	4,570	34.6	2,850	21.6
2030..........	5,420	37.6	4,970	34.5	3,160	21.9
Medium projections						
1990..........	4,240	42.7	3,180	32.0	1,670	16.8
2000..........	4,700	40.7	3,810	33.0	2,190	19.0
2010..........	5,190	39.2	4,490	33.9	2,740	20.6
2020..........	5,670	38.4	5,050	34.2	3,190	21.6
2030.:........	6,230	37.6	5,630	34.0	3,630	21.9
High projections						
1990..........	4,390	42.7	3,280	31.8	1,730	16.8
2000..........	5,010	40.7	4,020	32.7	2,330	19.0
2010..........	5,680	39.2	4,850	33.5	2,990	20.6
2020..........	6,370	38.4	5,610	33.8	3,590	21.6
2030..........	7,170	37.6	6,390	33.5	4,170	21.9

[1]Hardboard, insulating board, and particleboard (including waferboard, flakeboard, composite board, and medium-density fiberboard).
[a]1972 dollars. Use per 1,000 dollars of construction expenditure 1962–76 computed by Forest Service. (See table 3.11 for construction expenditures.)

Note: Data may not add to totals because of rounding.

Source: Forest Service, U.S. Department of Agriculture, An Analysis of the Timber Situation in tne United States 1952 – 2030.   1982

Exhibit 2.g

Basic Economic Indicators and Projections

Year	Population		Gross national product		Per capita gross national product		Disposable personal income		Per capita disposable personal income	
	Millions	Annual rate of change	Billion 1972 dollars	Annual rate of change	1972 dollars	Annual rate of change	Billion 1972 dollars	Annual rate of change	1972 dollars	Annual rate of change
1929.......	121.8	....	314.6	....	2,583	....	229.8	....	1,886	....
1933.......	125.7	0.8	222.1	−8.3	1,767	−9.1	169.7	−7.3	1,350	−8.0
1940.......	132.6	.8	343.3	6.4	2,589	5.6	244.3	5.3	1,849	4.6
1945.......	140.5	1.2	560.0	10.3	3,986	9.0	338.6	6.7	2,420	5.5
1950.......	152.3	1.6	533.5	−1.0	3,503	−2.6	361.9	1.3	2,386	− .3
1955.......	165.9	1.7	654.8	4.2	3,947	2.4	425.9	3.3	2,577	1.6
1960.......	180.7	1.7	736.8	2.4	4,077	.7	487.3	2.7	2,697	.9
1965.......	194.3	1.5	925.9	4.7	4,765	3.2	612.4	4.7	3,152	3.2
1970.......	204.9	1.1	1,075.3	3.0	5,248	1.9	741.6	3.9	3,619	2.8
1971.......	207.1	1.1	1,107.5	3.0	5,348	1.9	769.0	3.7	3,714	2.6
1972.......	208.8	.8	1,171.1	5.7	5,609	4.9	801.3	4.2	3,837	3.3
1973.......	210.4	.8	1,235.0	5.5	5,870	4.7	854.7	6.7	4,062	5.9
1974.......	211.9	.7	1,217.8	−1.4	5,747	−2.1	842.0	−1.5	3,973	−2.2
1975.......	213.6	.8	1,202.3	−1.3	5,629	−2.1	859.7	2.1	4,025	1.3
1976.......	215.2	.7	1,273.0	5.9	5,915	5.1	891.8	3.7	4,144	3.0
1977.......	216.9	.8	1,340.5	5.3	6,180	4.5	929.5	4.2	4,285	3.4
1978.......	218.7	.8	1,399.2	4.4	6,398	3.5	972.5	4.6	4,447	3.8
1979¹	220.6	.9	1,431.6	2.3	6,490	1.4	995.3	2.3	4,512	1.5
Low projections										
1990.......	236.3	.7	1,940	3.2	8,210	2.5	1,360	3.2	5,760	2.5
2000.......	245.9	.4	2,410	2.2	9,800	1.8	1,690	2.2	6,870	1.8
2010.......	250.9	.2	2,940	2.0	11,720	1.8	2,060	2.0	8,210	1.8
2020.......	253.0	.1	3,410	1.5	13,480	1.4	2,390	1.5	9,450	1.4
2030.......	249.3	−.1	4,000	1.6	16,040	1.8	2,800	1.6	11,230	1.7
Medium projections										
1990.......	243.5	.9	2,070	3.7	8,500	2.8	1,450	3.7	5,950	2.8
2000.......	260.4	.7	2,690	2.7	10,330	2.0	1,880	2.6	7,220	2.0
2010.......	275.3	.6	3,440	2.5	12,500	1.9	2,410	2.5	8,750	1.9
2020.......	290.1	.5	4,190	2.0	14,440	1.5	2,930	2.0	10,100	1.4
2030.......	300.3	.3	5,160	2.1	17,180	1.8	3,610	2.1	12,020	1.8
High projections										
1990.......	254.7	1.2	2,200	4.2	8,640	2.9	1,540	4.2	6,050	2.9
2000.......	282.8	1.1	3,010	3.2	10,640	2.1	2,110	3.2	7,460	2.1
2010.......	315.2	1.1	4,050	3.0	12,850	1.9	2,840	3.0	9,010	1.9
2020.......	354.1	1.2	5,180	2.5	14,630	1.3	3,630	2.5	10,250	1.3
2030.......	392.8	1.0	6,700	2.6	17,060	1.5	4,690	2.6	11,940	1.5

¹Preliminary.

Note: Annual rates of increase were calculated for the various periods indicated, except for the 1990 projections which were derived from the 1977 trend level ($1,290 billion) for gross national product.

Source: Forest Service, U.S. Department of Agriculture, An Analysis of the Timber Situation in the United States 1952 – 2030. 1982

3 Conventional Wood Adhesives

A large and growing share of structural wood products
are composed of pieces of wood held together by
adhesives. In a few cases, natural glues such as
animal glues, casein, and starch are still used. Paper
products and some types of low-density fiberboard rely
primarily on the natural adhesive qualities of wood
fibers. Currently, however, most wood-to-wood bonds in
composite wood panel products are produced by a variety
of synthetic resins -- primarily formaldehyde-based
thermosetting resins which were developed in the 1930s
and 1940s. (See Exhibit 3.a)

Any adhesive used in the bonding of wood must be able
to provide a strong, durable bond at a reasonable
price. Adhesive color, storage life, and ease of
application are also important factors. Because wood
is not a homogeneous substance, the resins used as wood
adhesives also must be effective over a wide range of
bonding conditions. Important wood quality variables
such as moisture content, density, and surface properties can vary considerably among different types of
wood and wood pieces. To adjust for such changes, a
number of different specific resin formulations are
required.

There is currently no universal resin that meets the
optimum requirements of all variables. Among the major
wood adhesive resins currently used, the main tradeoff
is between price and durability -- the cheapest resins
form the least durable bonds, and vice versa.

WOOD ADHESION

The creation of a bond between an adhesive resin and
the wood substrate requires adequate interpenetration
of the resin and wood components, and the development
of links between the resin and the exposed wood surfaces. However, even though the technology of
producing and applying wood adhesives is relatively
well-developed, the exact mechanisms that create the
actual bond between the wood substrate and the resin
are still not completely known.

Exhibit 3.a

U.S. Consumption of Natural and Synthetic Adhesives

	1971		1979		1985[1]	
	billion lb	%	billion lb	%	billion lb	%
Natural	1.25	45	1.2	30	1.3	24
Synthetic	1.55	55	2.8	70	4.2	76
Total	**2.8**	**100**	**4.0**	**100**	**5.5**	**100**

1 Estimated

Source: Arthur D. Little, Inc.

Adequate resin-wood contact in the adhesive application process is a necessary prerequisite to producing a good adhesive bond. The basic resin-wood contact is made in three phases: penetration, wetting, and diffusion. The liquid resin adhesive is applied to the wood; for wood flakes or particles, the resin is atomized and sprayed onto the surface, while for veneers, the resin is roll-coated as a film onto the surface. At a minimum, the resin must penetrate the wood surface to the level of solidly attached wood fibers without destroying the continuity of the resin remaining in the adhesive film layer. Wetting must then occur between the adhesive and solidly attached wood fibers, -- the surface tension of the adhesive film must be low enough to allow easy contact between the adhesive and wood molecules. Then the lower molecular weight fractions of the adhesive must diffuse (migrate) sufficiently into the wood fiber cell walls to establish a molecular "hold" on a wood constituent. This diffusion is accomplished by some combination of solvent (water, alcohol), heat, and/or pressure.[125]

The exact nature of the attractive forces between the resin and wood molecules, which produce the actual adhesive bond, has not been clearly determined. In general, it is thought that when resin molecules are brought into contact with wood molecules, a bond forms between reactive polar molecular groups of the resin and wood molecules. Three basic chemical processes are involved in adhesive bonding:

1) Van der Waals (dispersion) forces -- inter-molecular forces caused by the creation of temporary positive and negative electrical fields (polarities) in adjacent nonpolar molecules.

2) Hydrogen bonds -- electrostatic attraction between the hydrogen atom of one molecule and the flourine, oxygen, or nitrogen atom of another. These particular intermolecular bonds are an especially strong type of dipole attraction between polar molecules (molecules with a separation of positive and negative charges). In adhesives terminology, hydrogen bonding is called "specific adhesion."

3) Covalent bonds -- electrons are shared between two atoms; in wood adhesion, the formation of covalent

bonds between parts of resins molecules and wood
fiber molecules.[125]

Mechanical forces have also been hypothesized, but are
usually not considered a major contributor to the wood
adhesive bond. Studies have proven that all three
types of bonds are possible in wood-resin adhesion, but
the existence and/or relative importance of these
bonding forces in final wood bonds is not yet known.

Since wood-resin adhesion is essentially chemical in
nature, it is important that the resin contain
sufficient numbers of reactive groups and that the wood
surface present a significant number of reactive sites
to enable bonding. Any factors that limit resin
functionality or block reactive sites on the wood
structure necessarily impede adhesion.

SYNTHETIC THERMOSETTING WOOD ADHESIVE RESINS

Almost all of the adhesives currently used to
manufacture composite wood panel products are synthetic
polymer resins based on the condensation reaction of
formaldehyde with compounds derived from either benzene
(i.e., phenol, resorcinol) or ammonia (i.e., urea,
melamine). The four thermosetting resins produced from
these reactions are phenol-formaldehyde (PF), urea-
formaldehyde (UF), melamine-formaldehyde (MF), and
resorcinol-formaldehyde (RF) (See Figure 3.b) These
resins are termed "thermosetting" even though they do
not all require heat to initiate reaction; the term
"thermosetting" describes their insoluble and infusible
crosslinked compositions, typically achieved upon
heating. These resins have been found to offer the
best combination of cost and adhesion quality for
producing bonded wood products. After application to
the wood surfaces, these resins can be quickly modified
into hard, stable solids that adhere well to wood
surfaces, forming an interconnection between the pieces
of wood that is at least as strong as the wood itself.

All of these wood adhesive resins are formed by a
process of condensation polymerization. The two major
types of polymerization processes are addition and
condensation polymerization. In addition
polymerization, the polymer is formed without the loss
(splitting off) of any atoms. In the condensation
process, the polymer growth is accompanied by the

Exhibit 3.b

Thermosetting Resins: Flowsheet

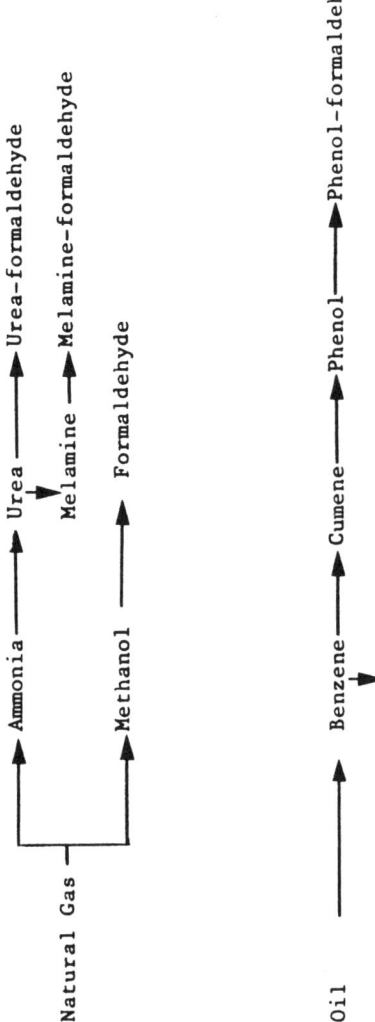

Source: Hagler, Bailly & Company

elimination of small molecules, often water or an alcohol.

One of the most important factors that influences the size and structure of polymer molecules is the number of reactive groups (atoms or groups of atoms that can form dipolar or covalent bonds) in the original monomer. If the original monomer has only two reactive groups, the resulting polymers are linear. If a monomer has three or more reactive groups, a branched polymer results, because two of the reactive groups participate in one polymer chain and the other(s) start the growth of one or more additional chains having one end attached to the original polymer chain. Under appropriate conditions, the branched chains can then form links with other linear chains and the polymer becomes "cross-linked."

All wood adhesive resins condense to become highly cross-linked polymers, although the details of their polymerization reactions and the exact structure of the resulting resin solids are not completely known. Simplified condensation and initial polymerization reactions for the four major types of wood panel resins are shown in Exhibit 3.c. The four basic monomers that are reacted with the formaldehyde contain 3-6 reactive groups:

phenol	3 groups + 1 polar hydroxyl (OH) group
resorcinol	3 groups + 2 hydroxyl groups
urea	4 groups (2 in each NH_2 group)
melamine	6 groups (2 in each NH_2 group)

In all of the reactions, the initial combination of formaldehyde and co-reactant is an addition polymerization, immediately followed by a condensation reaction of the intermediate products. Most of the resulting polymer bonds seem to be on methylene (CH2) "bridges" as the main cross-linking bonds.

In commercial production processes, the complete condensation reaction takes place in stages. In the initial manufacturing operation, the reaction is stopped at an early or intermediate stage, when the resin is still soluble in water or alcohol. This

Exhibit 3.c

Thermosetting Resin Addition and Condensation Polymerization

1. Phenol formaldehyde, addition step:

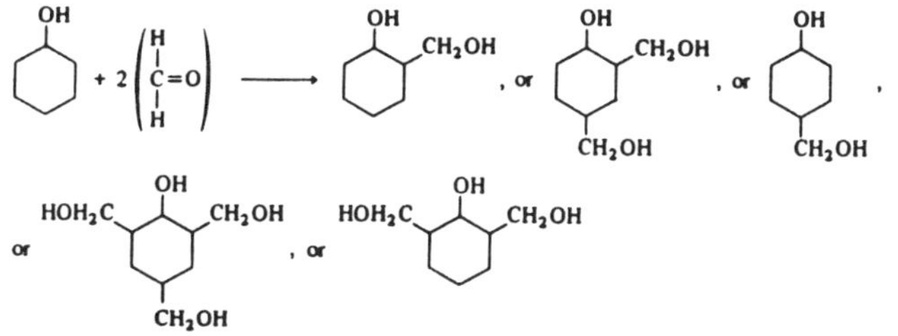

Condensation step:

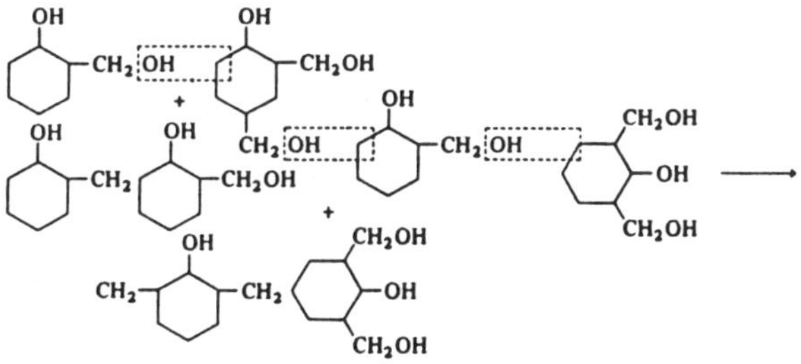

2. Resorcinol formaldehyde resin, addition phase:

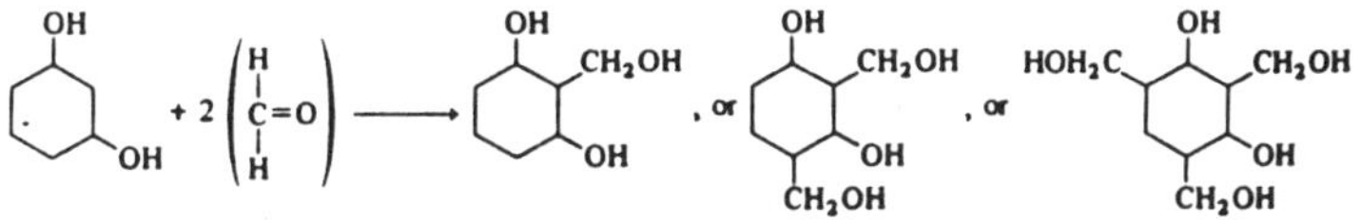

3. Urea formaldehyde resin; addition phase:

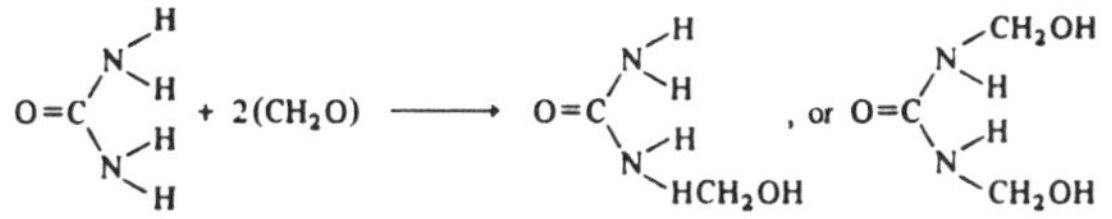

Condensation phase:

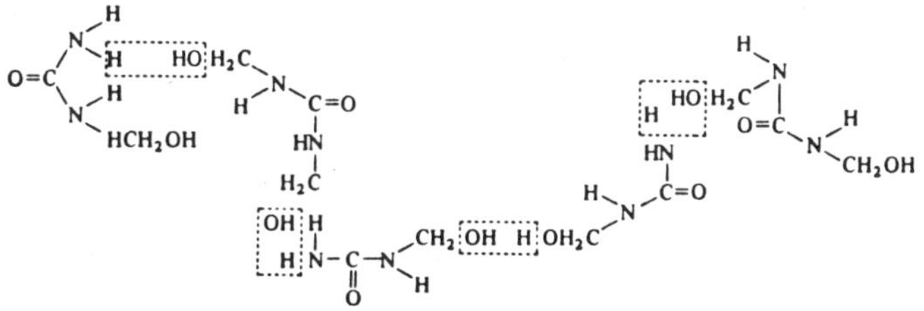

4. Melamine formaldehyde resin; addition phase:

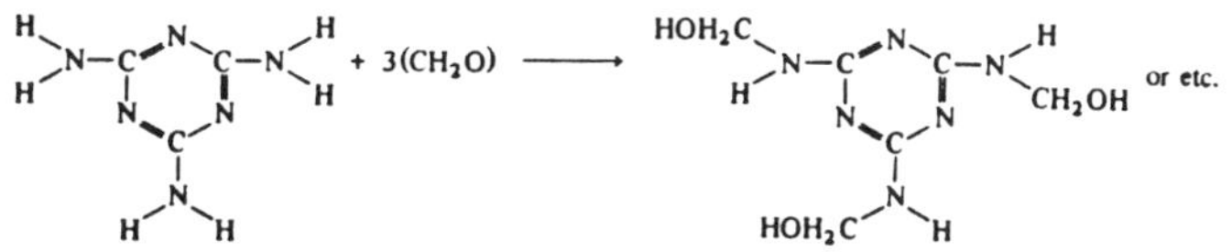

Source: Skeist, I. <u>Handbook of Adhesives</u>, 1984.

intermediate reaction product is then supplied to the
wood panel producer. At the point of final use, the
resin product, usually in a liquid dispersion, is
applied to the wood surface and the reaction is
restarted by adding some combination of heat, catalyst,
and/or additional formaldehyde. This additional
reaction, known as "curing" the resin, produces the
solid resin adhesive that joins the wood pieces into
panel products.

Wood adhesive resins are formed by combining the
appropriate molar concentrations of the two reactants
with a catalyst under specified temperature and
pressure conditions. By varying the chemical
ingredients and reaction variables, resins with a wide
range of properties can be formulated. In the basic
condensation reaction, the polymer molecules produced
by the reaction grow as water splits off at random
between reactive groups of neighboring molecules. As
the molecules grow, their properties change
continuously and the characteristics of the reaction
products differ considerably between lower and higher
condensation stages. Therefore, by carefully
controlling the reaction process and stopping it at the
proper time, resins with the necessary solubility,
viscosity, water retention, and curing rate are
produced.

Adhesive resins are almost always mixed with some
amounts of fillers, extenders, and other additives.
Fillers are added to provide bulk to the adhesive mix,
to control viscosity, and, in many cases, to help
absorb the water released in the final curing reaction.
Fillers are added in powdered form, and are usually
inert cellulosic materials made from wood wastes or
agricultural residues. An extender is an additive that
generally has some adhesive action -- either modifying
or enhancing the bonding effect of the resin adhesive
mixture. The use of extenders allows a reduction in
resin requirements, which therefore reduces adhesive
costs. Extenders are usually grain flours,
particularly soft wheat flours. (Although fillers and
extenders have significantly different properties, the
terms are often used interchangeably.) Other additives
that are sometimes incorporated into the adhesive
mixture include fire retardants, insecticides, and
compounds to aid in applying the adhesive mixture.[125]

There are over 20 major categories of wood products
that require the application of hydrocarbon-based
adhesives (see Exhibit 3.d for listing of wood product
categories and adhesive types). These 20 wood
categories consume roughly 40 percent of all synthetic
thermosetting resins consumed in the U.S. (see Exhibit
3.e). Together, phenol- and urea-formaldehyde resins
are the basis of nearly all industrial wood adhesives.

The following sections describe in more detail the four
major thermosetting resins used in the production of
composite wood panels:

- phenol-formaldehyde
- resorcinol-formaldehyde
- urea-formaldehyde
- melamine-formaldehyde.

All four resins are available in a wide range of
formulations to satisfy different operating conditions
and product requirements. In many cases, specific
adhesive resins and formulations are developed or
prescribed by adhesive manufacturers to meet individual
plant conditions and requirements.

Phenol-Formaldehyde Resins

Phenol-formaldehyde (PF) resins are the products of the
condensation reaction between phenol, a derivative of
benzene, and formaldehyde, a derivative of methanol.
Approximately 22 percent of total U.S. formaldehyde
supplies and 45 percent of phenol supplies are used in
the production of phenolic resins. Discovered by
Baekeland in 1907 and trademarked "Bakelite", PF resins
are characterized by their high strength, resistance to
moisture, dimensional stability, and low cost. (See
Exhibit 3.f)

Manufacture

A typical phenolic resin is made by a batch process in
a jacketed, stainless steel reaction kettle, equipped
with turbine-blade or anchor-type agitator and
condenser. (See Exhibit 3.g) Molten phenol and
formalin (37-40 percent aqueous solution of
formaldehyde) are charged to the kettle and agitation
is begun. The catalyst is then added and steam heat is
applied to the jacket to heat the mixture. Process

Exhibit 3.d

Wood Products Requiring Synthetic Resins Derived from Petrochemicals or Natural Gas

Wood Product Type	Synthetic Resin or Petrochemical
Softwood Plywood	Phenol-formaldehyde Adhesives
Hardwood Plywood	Phenol-formaldehyde, Urea-formaldehyde, Melamine-formaldehyde Adhesives
Laminated Beams	Phenol-formaldehyde, Resorcinol-formaldehyde Adhesives
Flooring & Decking	Phenol-formaldehyde, Melamine-formaldehyde, Resorcinol-formaldehyde Adhesives
Treated Lumber & Poles	Pentachlorophenol, Cresosote, Diesel Oil
Underlayment Particleboard	Urea-formaldehyde Binders, Wax Emulsions
Industrial Particleboard	Urea-formaldehyde Binders, Wax Emulsions, Polyester Fillers, Phenol-formaldehyde Binders, Urethane/Isocyanate Binders
Medium Density Fiberboard	Urea-formaldehyde, Melamine-formaldehyde Binders, Wax Emulsions
Wet-formed Hardboard	Phenol-formaldehyde Binders, Wax Emulsions
Dry-formed Hardboard	Phenol-formaldehyde Binders, Wax Emulsions
Wet-formed Insulation Board	Urea-formaldehyde, Phenol-formaldehyde Binders, Wax Emulsions
Structural Panelboard	Phenol-formaldehyde Binders & Adhesives, Wax Emulsions
Paper Overlays	Polyester, Phenol-formaldehyde, Melamine-formaldehyde & Urea-formaldehyde Saturants
Low Pressure Decorative Laminates	Melamine-formaldehyde Saturants
High Pressure Decorative Laminates	Phenol-formaldehyde, Melamine-formaldehyde Saturants
Paperboard Products	Cationic Urea-formaldehyde & Polyamide Wet Strength Agents, Styrene/Acrylic Emulsion Sizing Agents
Paper Toweling	Cationic Melamine-formaldehyde Agents
Paper Coatings	Melamine-formaldehyde, Urea-formaldehyde, Styrene/Acrylic Emulsion Binders, Hydrocarbon Resins, Phenolics, Vinyl Acetate Emulsions
Furniture Finishes	Alkyd/Urea-formaldehyde & Melamine-formaldehyde Base Coats; Water Based Acrylics Base Coats; Alkyd/Amine, Alkyd/Acrylic Topcoats; Alkyd/Urea-formaldehyde Topcoats
Oil & Air Filter Paper Products	Melamine-formaldehyde and Phenol-formaldehyde Saturants
Paper Tubing Products	Phenol-formaldehyde Saturants
Molded Wood Products	Phenol-formaldehyde, Urea-formaldehyde, and Melamine-formaldehyde Binders

Source: Gillespie, Robert H. ed. Adhesives for Wood, 1984.

Exhibit 3.e

U.S. Production of Thermosetting Resins (1983)

| Resin | Production (million lb, dry weight) | | % Used in Wood Products |
	Total	For Wood Panel Products	
Resorcinol-formalde-hyde	31	3	11
Urea-formaldehyde	1,172	880	75
Melamine-formaldehyde	180	11	6
Phenol-formaldehyde	1,320	594	45
Totals	2,703	1,488	55

Source: U.S. Department of Commerce

Exhibit 3.f

Phenol-Formaldehyde Adhesives: General Characterization

Type	Synthetic thermosetting resin
Physical form	Spray-dried powder to be mixed with water; alcohol, acetone, or water solvent solutions; glue film. May be compounded with fillers and extenders.
Shelf life (@20°C)	Powders and resins: 1-2 yr Film: up to 1 yr
Working life (@20°C)	Powders and resins: 1-24 hr Film: indefinite
Closed assembly time (@20°C)	24 hr or more depending on wood type
Setting process	Condensation polymerization with elimination of water
Processing conditions	Resins, up to 15 min at 100-150°C and 700-1700 kPa bonding pressure. Film, up to 15 min at 120-150°C and 700-1400 kPa.
Coverage	0.05-0.075 kg/m^2
Permanence	Resistant to weather, boiling water, and biodeterioration.
Applications	Fabrication of exterior-grade weather and boilproof plywood. Glass to metal adhesive for electric light bulbs.
Comments	Gap-filling properties are poor and inferior to acid catalysed phenolic resins. Joints require conditioning period of up to 2 days. Though durable and resistant to many solvents, bonds are brittle and prone to fracture under vibration and sudden impact. Alkaline nature of adhesive precludes the use of brass or copper mixing vessels. Phenol-formaldehyde is employed as an additive to other materials to form adhesives for glass and metals; as a modifying agent for thermoplastic elastomer adhesives or as a component of thermoplastic resin-elastomer adhesives for metal bonding.

Source: Shields, J. Adhesives Handbook, 1984.

Exhibit 3.g

Phenol-Formaldehyde Production Unit

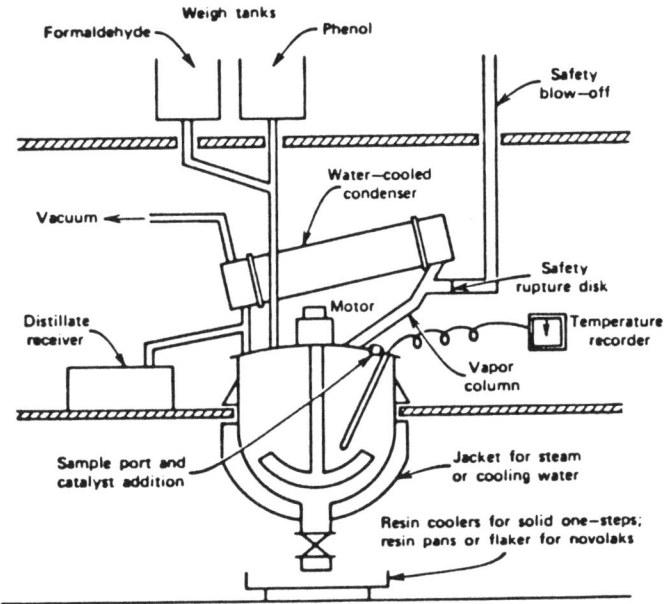

Source: Kirk-Othmer Encyclopedia of Chemical Technology, 1985

temperature and reaction time depends on pH and the phenol:formaldehyde molar ratio.

Following the reaction period, the resin batch is dehydrated and processed into its final form, which can be either a solid (lump, powder, or flake) or a water or alcohol solution. In general, the solid forms have a longer shelf life than the liquids.

Although the basic manufacturing process can produce an infinite variety of phenolic resins, all of the commercially important products fall into two main categories: one-step (resoles) and two-step (novolacs). The differences between these two basic types of phenolic resins are determined by differences in the phenol:formaldehyde molar ratio, the type and amount of catalyst used, and the time and temperature of reaction.

The one-step resins (resoles) are manufactured by using an alkaline catalyst, such as sodium hydroxide, and a phenol-to-formaldehyde molar ratio in excess of 1:1. The reaction temperature is 175-212°F and the reaction time is 1-3 hours. Resole resins have reactive methylol side and end groups that provide the necessary reactive sites for final crosslinking with the wood surface upon final curing. To cure an alkaline-catalyzed resin requires no additional reactants, only an increase in temperature. Since the cure is merely the continuation of the original reaction, the resole is commonly refered to as a one-step resin.

Two-step phenolics (novalacs) are produced from a reaction that uses an acidic catalyst (sulfuric acid) and a phenol-to-formaldehyde ratio less than 1:1. The reaction temperature is 212°F, and the reaction time is 3-6 hours. These novolac resins have a low number of reactive methylol side chains, thus restricting crosslinking capabilities but increasing shelf life. To provide the additional reactive sites needed for suitable crosslinking with the wood surface, additional formaldehyde in the form of hexamethylene tetramine (or "hexa") is added to complete the condensation reaction in the curing phase. Because of this extra processing step required for novalac resins, the one-step resole resins are almost exclusively used in the wood products industry.[4]

Market

Phenol-formaldehyde resins have a wide variety of uses.
In 1983, U.S. production of phenol-formaldehyde resins
totaled 1320 million pounds of dry weight resin. Of
this production, approximately 65 percent was consumed
by wood products such as plywood and laminates. Of
this, exterior grade plywood was the leading consumer
at over 52 percent because of its requirement for the
superior weather resistant qualities which phenolic
adhesives offer. Other applications for phenol-
formaldehyde resins are in molding compounds,
insulation materials, abrasives, coatings, and foundary
and shell moldings (see Exhibit 3.h).[23]

In 1983, phenol-formaldehyde resins were produced by
over 20 U.S. companies. However, Borden Chemical,
Georgia Pacific, Chembond, and Reichhold were the major
players in this market. Over 30 companies formulated
these phenolic resins into phenol-formaldehyde and
phenol-resorcinol-formaldehyde resins (see Appendix A
for list of formulators). Since many of these
companies are small and regional in nature, serving
local wood products companies, concentrations of
phenolic resin plants are located in areas close to the
wood products industry such as the Northwest (i.e.,
California, Oregon, Washington) and the Southeast
(i.e., Alabama, Georgia, North Carolina). Plants are
also located in the Michigan area to supply phenolic
resins to the automotive industry.

Only a small number of phenolic adhesive formulators
are major chemical companies. The rest of the
producers are either small chemical companies or
companies that derive their principal income from
outside the chemical industry but use the resins
captively, such as lumber and paper, fiberglass, and
paint manufacturers. Often the adhesive formulators
are not themselves resin manufacturers, but buy the
resins from domestic or foreign commodity chemical
producers.[142, 143]

Less than 1 percent of the U.S. production of phenolic
resins was exported in 1983. The major market for U.S.
PF resin exports was Canada. Almost 3 percent of U.S.
phenolic resin consumption was imported in this same
year. The major sources for these imports were Canada,
Switzerland, and the Netherlands. Over the 1979 to
1983 time period, U.S. exports declined by over 46

Exhibit 3.h

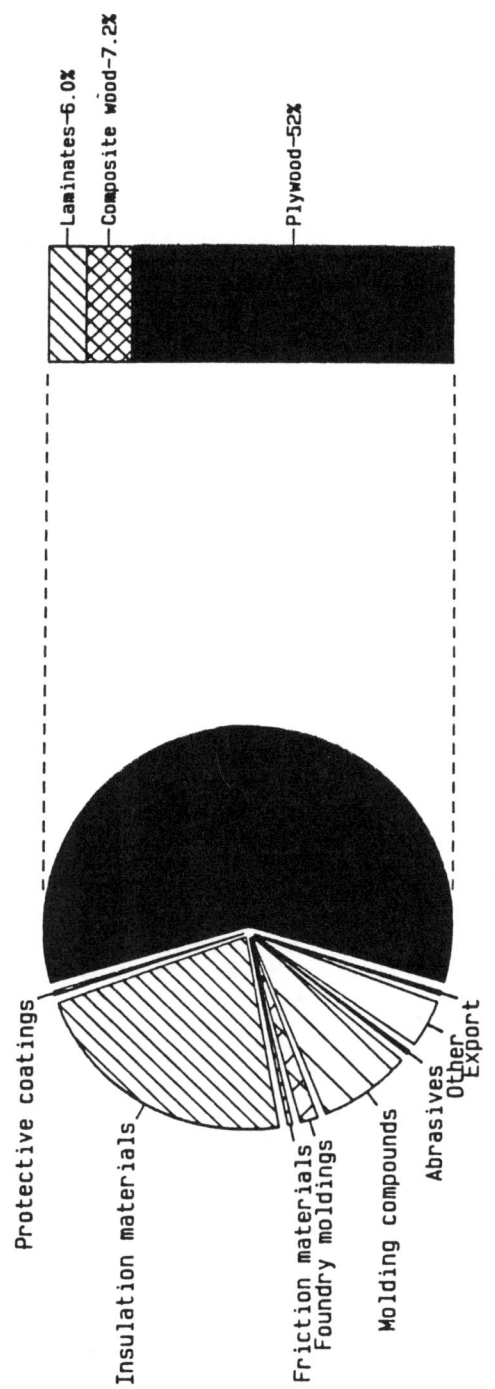

U.S. Phenol-Formaldehyde
Resin Market Applications
(% of 1983 Sales)

Laminates-6.0%

Composite wood-7.2%

Plywood-52%

WOOD ADHESIVES - 65.2 %

Protective coatings

Insulation materials

Friction materials
Foundry moldings

Molding compounds

Abrasives
Other
Export

Source: Modern Plastics, 1984

percent, while imports increased by over 451 percent
during this same time frame.

Wood Adhesive Application Processes

Almost all phenol-formaldehyde adhesives use one-step
resole resins. Phenolic resins can be used in liquid
or powdered form, but the liquid form predominates in
wood adhesive applications.

In the typical hot-press plywood or particleboard
manufacturing operation, a one-step phenolic resin is
applied as a liquid in the form of 25-50 percent (by
weight) resin solids dispersed in a highly alkaline
water system. Before application to the wood and final
curing, fillers and extenders (in powdered form) are
added to the liquid to optimize adhesive qualities and
to reduce costs. The specific mix of ingredients for
applying phenolic adhesives varies according to the
product and the manufacturing process involved. The
relative proportions (by weight) of ingredients in a
typical phenolic resin adhesive mixture for making
southern pine plywood -- the largest use of phenolic
resins -- is:

- adhesive resin solids   28%
- filler   8%
- extender   5%
- dry sodium hydroxide   1.5%
- water   57.5%

The adhesive resin solids share includes 3-5 percent
sodium hydroxide. The filler and extender also contain
water (10-15 percent total for both ingredients).

After the adhesive is applied to the wood substrate,
the veneers (in plywood) or chips (in particleboard)
are assembled and then cured in heated presses. In
plywood manufacture, the curing time is approximately 5
minutes at 270-300°F and 200-500 psi.[119]

Resorcinol-Formaldehyde Resins

Resorcinol-formaldehyde (RF) resins are the products of
the condensation reaction between resorcinol (also
known as m-dihydroxybenzene) and formaldehyde (see
Exhibit 3.c). RF resins used for adhesives are
prepared by reacting resorcinol with deficient amounts
of formaldehyde. At the time of use, additional

formaldehyde is added, and final curing to the solid state takes place.

RF resins are the strongest and most durable of the four types of thermosetting wood adhesive resins. They are often described as the "ideal adhesive" because they offer durable, waterproof bonding and cure at ambient temperatures in a relatively short time.

The main disadvantages of RF resins are high cost and a dark color, which limits the use of RF in wood products to applications that require exceptionally high strength and durability and where dark gluelines are not a problem.   Therefore, its use has been limited to marine wood applications or to applications where it acts as a copolymer with phenol-formaldehyde resins in thick lamination applications, in which the process energy savings from resorcinol's greater reactivity at cool or moderately warm curing temperatures outweighs its high costs.   (See Exhibit 3.i)

Manufacture

RF resins are produced from condensation of resorcinol (derived from benzene) with formaldehyde (derived from methanol).   The process requires no catalyst.   The condensation of resorcinol and formaldehyde is carried out in equipment similar to that used in producing phenolic resins.   A wide range of RF resin formulations are possible, depending on the specific combination of molar ratios, concentration, temperature, and pH. Because the basic reaction is very rapid and exothermic, an alcohol -- usually methanol -- is added to control the speed of the reaction.   Alcohols perform other useful functions during the application of the resin as an adhesive, so they are left in the final liquid product.

Since resorcinol is so highly reactive, only part of the formaldehyde needed to produce a fully-cured RF resin is introduced into the initial resorcinol-formaldehyde condensation reaction.   RF resins are similar to phenolic resins in that they readily combine with formaldehyde to form methylol derivatives, whose reactivity is so high that they continue to react under ambient, uncatalyzed conditions with the formaldehyde, resorcinol, or other methylol-containing molecules to form high molecular weight polymer chains.   The superior reactivity of RF resins compared to that of

Exhibit 3.1

Resorcinol-Formaldehyde and Phenol-Resorcinol-Formaldehyde

Type	Synthetic thermosetting resins. Phenol-resorcinolformaldehyde is a copolymer of resorcinol-formaldehyde and phenol-formaldehyde.
Physical form	Red-brown resin and hardening agent (liquid or powder). Resin may be diluted with alcohol or water or extended with filler.
Shelf life (@20°C)	Resin and liquid hardener: 1 yr Powder: indefinite
Working life (@20°C)	3-4 hrs (1 hr @ 38°C)
Closed assembly time (@20°C)	1/2 - 2 hrs
Setting process	Condensation polymerization with elimination of water.
Processing conditions	For general purposes, 4-8 min at 100°C or 8-10 hr at 20°C and 350-1000 kPa bonding pressure. Timber (Hardwood) 8-15 hr at 20-40°C and 1000kPa; (Softwood) 8-15 hr at 20-40°C and 700 kPa. Moisture content of wood should be below 15%. Plastics, cold cured or heat cured below softening point of plastic. Contact to 350 kPa bonding pressure. Press time decreases with increasing temperature; some assemblies bonded at 45°C require 1 hr press time, those glued at 90°C require 3 min press time. Resins set with difficulty under winter temperature conditions.
Coverage	general purpose bonding and plastics: 05-1 kg/m^2 timber construction work: 3.2-6.25 kg/m^2 plywood fabrication: 3.05-3.125 kg/m^2
Permanence	Durability of bond is generally comparable to that of heat setting phenolic adhesives and limited by the durability of the adherend material. Resistance to severe weather humidity, boiling water, temperature, and biodeterioration is excellent. Maximum chemical resistnce after 6 days. Maximum durability for hardwood assemblies is achieved by curing at moderately high temperatures.
Applications	Plywood and timber structures for exterior use; ship and marine constructions; wooden structural assemblies such as roofs, bridges, and frameworks; wood to metal laminates where the metal is primed with metal bonding adhesives (epoxies). Bonding of acrylic, polyamide, molded urea and phenolic plastics.
Comments	Bond strengths for wood usually exceed the strength of this material; neutral character of resorcinol resins avoids damage to cellulosic fibers. Long storage life, and non-reactivity until catalyst addition is made, are valuable properties. Hot-cured assemblies should not be handled or machined for at least 24 hr. Low temperature cured bonds require a 7 day conditioning period. Gap-filling properties are improved by filler additives; phenol-resorcinols are suitable for glue-lines up to 0.4 mm thick. Staining power of resins often precludes bonding light colored woods and veneers. Although phenol-resorcinol resins cost less than resorcinol resins, both types are more expensive than the urea resins used for similar applications. Long storage life, and non-rectivity until catalyst addition is made, are valuable properties.

Source: Shields, J. <u>Adhesives Handbook</u>, 1984.

phenolics is due to the presence of two reactive
hydroxyl groups rather than one.

A resorcinol-to-formaldehyde molar ratio of 1:0.5-0.7
yields a stable (uncured) RF resin; more formaldehyde
at this stage produces an unstable or gelatinous
product. For final curing, a "hardener" containing
additional formaldehyde is added, producing the final
solid resin product. Neither a catalyst nor heat are
required for final curing, although curing under mildly
alkaline conditions is helpful in some applications and
heat is often used to speed the curing process.

RF-based adhesives are sold as two-part systems. The
resin is provided in a liquid form, with about 50
percent resin solids in an alcohol/water solution. The
formaldehyde source, or hardener, is in powder form.
Paraformaldehyde is the most widely-used hardening
agent. Fillers may also be included in the hardener
mixture.

Due to its high cost, most of the resorcinol used in
wood products is currently provided in the form of
resorcinol-phenol (PF-RF) copolymer resins. In order
to produce these resins, phenol is combined with
formaldehyde before the resorcinol is introduced. If
the resorcinol were added to the initial charge, it
would preempt most of the formaldehyde and form a gel
because it is many times more reactive than phenol.
The PF-RF resin manufacturing process can be adjusted
to yield a wide variety of adhesive products. Such
combined resins generally contain about 15 percent PF
resin.[125]

Market

Consumption of resorcinol in the U.S. is distributed
among the following end use categories:

- rubber products   68%
- wood products   11%
- miscellaneous products   21%.

Most of the resorcinol is used in the rubber products
sector. In the miscellaneous category, resorcinol acts
as a chemical intermediate in such specialty products
as ultraviolet light absorbents, dyes, pharmaceuticals,
and agricultural chemicals. All resorcinol used in the
wood products sector is in the form of RF resins. In

the wood products market, the major use is in laminated beams and other timber structures, particularly those intended for exterior construction. The RF resins are especially well-suited for wood bonding in marine applications because of their superior waterproof bonds.

Annual domestic consumption of resorcinol is slightly under 20 million lbs. Only one company in the U.S., Koppers, produces resorcinol. According to Koppers, they are the world's largest resorcinol producer. U.S. annual production averages approximately 30 million lbs with a total capacity of 45 million lbs., resulting in an operating rate of roughly 67 percent. Approximately a third of this U.S. production is exported (see Exhibit 3.j).

The U.S. imports over 2 million lbs of resorcinol each year. The sources of these imports in 1983 were Japan (70 percent), and West Germany (30 percent). In 1985, West Germany is expected to claim a larger share of the import market.

Wood Adhesive Application Processes

Since RF resins are used in a variety of wood laminating operations, there is no single dominant application process. However, the two main considerations in most RF laminating operations are the short potlife of RF-based resins (once hardener is added) and the need to minimize adhesive wastage because of the high cost of the resin.

A typical two-part phenol-resorcinol (PF-RF) resin adhesive is mixed from a 5:1 mass ratio of resinous liquid (50-60 percent solids) and powder hardener (10 parts paraformaldehyde and 10 parts filler). Fillers are primarily wood or nut shell flours. RF-based resins normally do not contain extenders. Once the liquid and solid portions are mixed, adhesive application and product assembly must be quickly completed since such adhesives have a potlife of only 3-4 hours. Initial cure time is 8-15 hours at room temperatures.[125]

Urea-Formaldehyde Resins

Urea-formaldehyde (UF) resins are the products of the condensation reaction between urea and formaldehyde in

Exhibit 3.j

Resorcinol: Comparison of Major Producing Countries (1982)

	U.S.	Western Europe	Japan
Production (million lbs)	28	15-16	7
Capacity (million lbs)	40-45	26	18
Operating Rate (%)	62-70%	58-62%	39%
Domestic Demand (million lbs)	19	16	3
Domestic Demand by end-use:			
- rubber products	68%	50%	55%
- wood products	11%	25-30%	20%
- miscellaneous*	21%	20-25%	25%
Exports (million lbs)	10	4	4
Imports (million lbs)	2	6	1
Price (1984 $/lb)	$1.65-1.80	$1.50	$2.80-3.35

* Includes dyes, pharmaceuticals, ultraviolet absorbers, fungicidal creams, agricultural chemicals, and explosives

Source: Chemical Economy & Engineering Review; Chemical Marketing Reporter

the presence of an alkyline catalyst. Final curing of
UF resins requires the addition of an acid or acid-
releasing catalyst, which is usually combined with the
resin at the time of use. With sufficient amounts of
catalyst, UF resins can be cured at room temperature,
but, in most commercial uses, the resin is cured at
higher temperatures.

The main advantages of the UF resins are low cost,
versatility, low-temperature curing, and a colorless
glue line. Their main disadvantages are the need for a
two-component formulation, low moisture resistance, and
high formaldehyde emissions (see Exhibit 3.k).[85]
Combined urea/melamine-formaldehyde resins are often
used to improve durability.

Manufacture

UF resins are made from urea (derived from ammonia),
formaldehyde (derived from methanol), and an alkyline
catalyst. Approximately 30 percent of all U.S. formaldehyde supply and 9 percent of urea supply is used to
produce UF resins.

The basic UF production process uses a batch process,
with the reactants combined in a reaction kettle that
is equipped with an agitator and condenser and jacketed
or coiled for heating and cooling (See Exhibit 3.l).
The urea is charged to a reactor which contains
formalin (37-40 percent aqueous solution of formaldehyde). The urea may be added fully at the start of
the reaction or stepwise to obtain variations in
polymer distribution and reactivity of the final
product. Sodium hydroxide may be used for pH adjustment. Some of the water produced in the reaction is
removed by vacuum distillation.

Depending on the desired resin properties, the
manufacturing variables can vary considerably.
Temperatures can range from 175°-$250^{\circ}F$ and reaction
times range from 40 minutes to 24 hours. A typical
process combination for producing a resin for use in
particleboard -- the major application of UF resins
-- would include a formaldehyde-to-urea molar ratio of
1.5-1.3:1, a temperature of 195°-$212^{\circ}F$, pH of 8.5-9.0,
and a reaction time of 40 minutes.

Control of the reaction variables -- particularly the
molar ratios, pH values, and temperature -- is critical

Exhibit 3.k

Urea-Formaldehyde Adhesives: General Characterization

Type	Synthetic thermosetting resin
Physical form	Two part products consisting of the resin and hardening agent (liquid or powder). Also available as spray—dried powder, with incorporated hardener, which is activated by mixing with water.
Shelf life (@20°C)	Resins: 3-6 mth Powders: at least 1 yr. Catalysts: indefinite
Working life (@ 20°C)	From 0-48 hr according to composition and setting temperature. Room temperature setting types range from 3-10 hr
Closed assembly time (@20°C)	From 0-24 hr depending on composition. From 0-30 min where resin and catalyst are applied separately to each adherend.
Setting process	Condensation polymerization with elimination of water
Processing conditions	Suitable for use with radio-frequency equipment. For general purposes, 2-4 hr at 20°C and 350-700 kPa bonding pressure. For plywood manufacture, 5-10 min at 120°C and up to 1600 kPa. Timber: (Hardwood) 15-24 hr at 20°C and 1400 kPa. (Softwood) 15-24 hr at 20°C and 700 700 kPa. Bonding pressures depend on type of wood, shape of parts and similar factors.
Coverage	general purpose bonding: 0.1 kg/m^2 plywood fabrication: 0.075 kg/m^2 timber construction work: 0.125-0.2 kg/m^2
Permanence	Urea formaldehyde adhesives lack the durability of combinations with melamine, phenolic, or resorcinol resins, and are unsuitable for service conditions which are extreme, e.g. high humidity, boiling water, and temperatures above 60°C. Resistant to cold water and biodeterioration and may be compounded with additives to improve durability. Hardening agents used are important components and manufacturers advice should be followed. Resins cured with ammonium thiocyanate have superior hot water resistance to those hardened with diammonium phosphate. Use of starch extender reduces water resistance, increases resin viscosity to overcome adhesive absorption into porous materials (wood) and lowers cost of adhesive. Use of phenolic or resorcinol additives promotes resistance to water, weather and temperature changes; durability improved but increases cost of adhesive. Use of melamine fortifier increases resistance to water and temperature and improves durability. Melamine or resorcinol fortifiers do not enhance durability beyond that of phenolic resins.
Applications	Wood and related materials with moisture content of 7-15%. Manufacture of light woods, thin veneers, and plywood furniture assembly .and joinery. Timber construction subject to low stresses.
Comments	Development of full bond strength is gradual over a few days but assemblies may be handled after 6-12 hr. Bond strengths usually exceed the strength of the wood when glue-line thickness ranges from 0.05-0.16 mm. Gap-filling properties are poor where glue-lines exceed 0.37 mm. Thick glue films will craze and weaken bond strength unless special modifiers, such as furfural alcohol resins, are used. Gap-filling properties are important where the adhesive is extended with wood or walnut shell flow. Adhesives are non-staining and light colored; unsuitable for exterior conditions or extreme temperatures.

Source: Shields, J. Adhesives Handbook, 1984.

Exhibit 3.1

Urea- and Melamine-Formaldehyde Manufacture

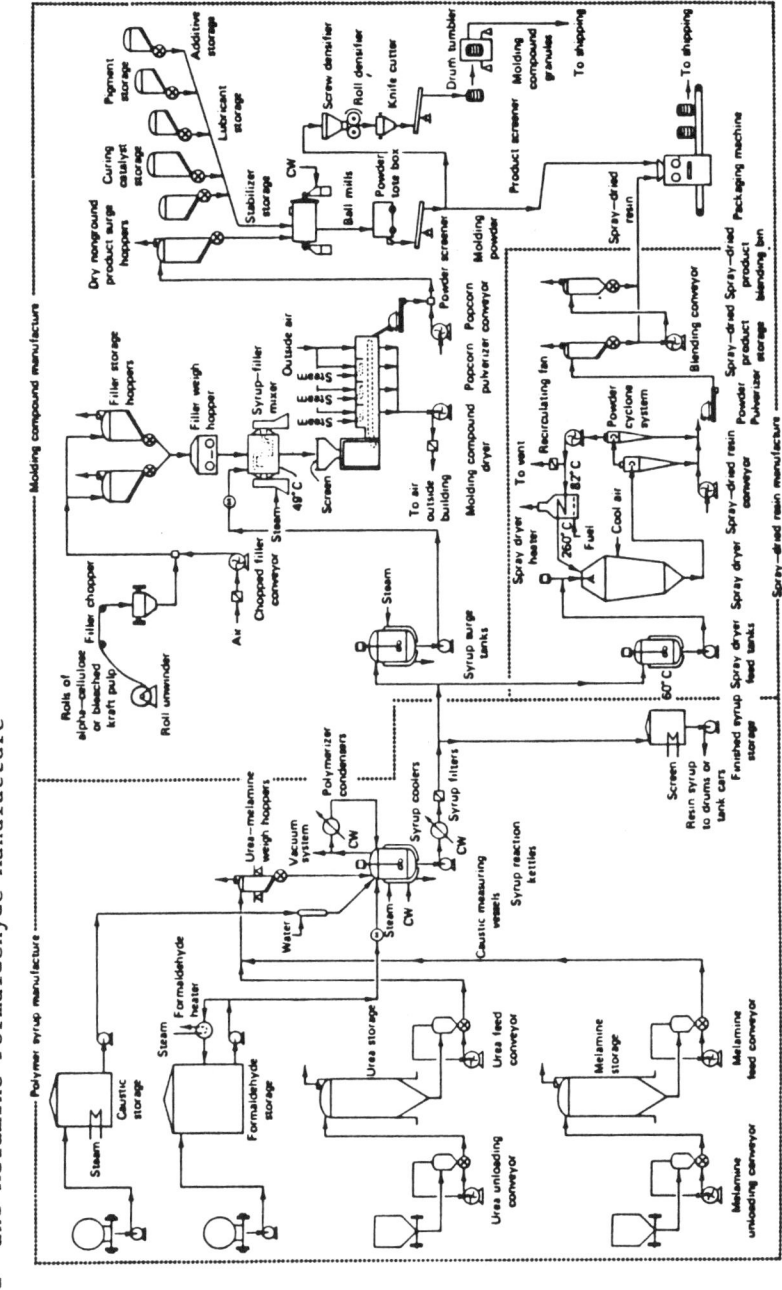

Source: Kirk-Othmer Concise Encyclopedia of Chemical Technology, 1985

to determining the final properties of the resin.   In
the basic condensation reaction, the polymer molecules
produced by the reaction grow as water splits off at
random between reactive groups of neighboring
molecules. As the molecules grow, their properties
change continuously and the characteristics of the
reaction products differ considerably between lower and
higher condensation stages. Therefore, by carefully
controlling the reaction process, and stopping it at
the proper time, UF resins with the necessary
solubility, viscosity, water retention, and curing rate
are produced. After sufficient reaction has occured to
yield the desired product qualities, the pH of the
resin is adjusted to halt the polymerization reaction
and the liquid product is cooled.

UF resins are generally marketed in liquid form (as
water suspensions) when large-scale use is involved and
shipping distances are not excessive. Liquid UF resins
are available with solids contents of about 40 to 70
percent. Some UF resins are spray-dried and are
marketed as powders with or without catalysts
incorporated.[3, 85, 125]

Market

Approximately 75 percent of U.S. urea-formaldehyde
resin supplies are used as adhesives in wood products
(see Exhibit 3.m). Uses in other areas include fiber
insulation binders and paperboard adhesives. In
general, UF resins are used in bonding products
intended for interior use, where only limited
resistance to heat and moisture is required.
Particleboard manufacture accounts for over 90 percent
of all UF adhesives used in the wood products sector,
with additional minor usage in interior plywood.
Various formulations of urea resins are also used
extensively in many types of furniture assembly
processes including laminating, end joint bonding, and
veneer splicing.

Under conditions of high temperatures and humidity,
particleboard made with currently available UF resins
releases formaldehyde into the air, which can cause
odor and health problems in enclosed interior spaces.
The formaldehyde release can be due to (1) free,
unreacted formaldehyde present in the board, and (2)
formaldehyde formed by gradual breakdown (hydrolysis)
of the urea resin bonds in the presence of moisture.

Exhibit 3.m

U.S. Urea and Melamine Resin Market Applications (% of 1983 Sales)

Plywood-3.3%
Laminates-2.8%
Composite wood-68.9%

WOOD ADHESIVES - 75 %

Export
Molding compounds
Paper treatment
Textile treatment
Protective coatings
Other

Source: Modern Plastics, 1984

In 1984, the U.S. Department of Housing and Urban
Development (HUD) set limits on formaldehyde emissions
from hardwood plywood and particleboard. Further
regulations are expected soon for medium-density
fiberboard.

The U.S. Environmental Protection Agency, which
recently announced stricter standards on formaldehyde
emissions in the workplace, is becoming increasingly
concerned about this release of formaldehyde from UF
resins. Future regulation limiting its use is likely,
possibly during the next 5 years.[141]

In 1983, the U.S. produced 1.2 billion lb (dry weight)
of UF resins. Major producers include Borden Chemical,
Georgia Pacific, and American Cyanamid. In 1983, the
U.S. exported over 5.8 million lbs of UF resins. The
primary destinations (in order of importance) were
Canada, Hong Kong, and the Netherlands. In this same
year, just over 15 million lbs of UF resins were
imported, with Israel, the United Kingdom, and Canada
as the major suppliers (in order of importance).

Since urea resin manufacturing costs are not
significantly affected by small plant size and the raw
materials are generally available, many developing
countries manufacture this resin, including Taiwan,
India, the Republic of Korea, and Malaysia.

Wood Adhesive Application Processes

In all important commercial applications, UF resins are
applied in liquid form, with a catalyst added to the
mixture shortly before application. The addition of an
acidic or acid-releasing catalyst (hardener) begins the
curing process by restarting the original condensation
reaction between the urea and formaldehyde. The
catalyst promotes additional condensations of the
monomers and polymers, which react to form larger
molecular structures -- releasing water in the process
-- until the solid, cured resin is formed.

A wide range of hardener catalysts are used, but most
formulations are based on solutions of ammonium salts,
generally ammonium chloride or ammonium sulfate, or
mixtures of ammonium chloride and urea. The amount of
hardener needed for proper curing varies between 1 and
5 percent (by weight) of the combined resin/catalyst
mixture.

As with the other types of resins, various fillers and/or extenders are also added to the adhesive mixture before final application to the wood surface. The most commonly used fillers and extenders are cellulosic flours, such as walnut flour, and cereal flours. As a general rule, 20-50 percent filler is used for high-strength uses, such as structural laminates and joints, and up to 100 percent (equal parts filler and resin solids) for plywood. In particleboard production, a small amount of wax is also added to the bonding mixture to enhance water resistance.

UF resins can be cured in either hot-press or cold-press operations. Heat is usually applied to reduce curing time and/or reduce the amount of catalyst required. The relative proportion of catalyst also affects curing variables; more catalyst results in lower curing times at a given temperature, or lower temperatures for a given curing time.

Most particleboard and plywood made with urea-formaldehyde resins are produced in hot-press operations, since short assembly time is a primary requirement. Cold-press operations are mainly limited to small-scale bonding operations.

In a typical particleboard hot-press operation, the wood particles are conditioned to a moisture content of 5-10 percent. The glue mix is generally composed of a liquid resin to which additional water has been added to facilitate spraying, plus small amounts of acidic catalyst. Small quantities of insecticides, wax emulsion, and fire-retarding agents are added before spraying the adhesive onto the wood chips. Fillers and/or extenders are also added. The amount of resin solids is usually 6-10 percent of the final dry weight of the board; the comparable share of wax is 1-2 percent.

After blending with the binder and other additives, the wood particles are formed into mats and then pressed between heated platens. Press temperatures range between 250-300°F. The curing time depends on the thickness of the board and other variables, but generally is 8-20 minutes. Pressures used are in the range of 285-500 psi.

Other wood panel hot-press applications are essentially similar to the particleboard operations. Some laminating operations use high-frequency heating (passing an electric current through the assembly) to speed cure time. Cold-press operations are carried out at room temperature (but at least 70°F) and with much longer cure times (several hours).[85, 104, 125]

Melamine-Formaldehyde Resins

Melamine, like urea, is an amino compound that is reacted with formaldehyde to form a thermosetting resin. In general, melamine-formaldehyde (MF) resins have significantly better application and bonding properties than urea-formaldehyde (UF) resins. MF resins are heat-cured without the addition of a catalyst, but they cannot be cured at room temperature, even with a catalyst (See Exhibit 3.n).

The major drawback of MF resins in wood adhesive uses is their high cost -- MF resins are more expensive than UF or phenolic resins. Consequently, MF resins are used only in wood products that require more durability and water resistance than UF resins can provide, or in specialty products which require the white glueline which MF resins provide. Because of the cost/quality tradeoff, various mixtures of MF and UF resins are often used.

Although MF resins are more resistant to heat and moisture than UF resins, bonds formed by MF resins are still subject to deterioration (through hydrolysis of the amino bonds) from prolonged exposure to heat and moisture.[62, 123]

Manufacture

The production process for melamine is more complex than for urea, and is the major reason for the large cost differential between MF and UF resins. The technical chemical name for melamine is 2,4,6-triamino-1,3,5-triazine. It is a cyclic compound containing nitrogen in the ring. It is currently produced from urea through several commercial processes.

The differences in MF and UF properties and catalyst requirements are based on the fact that melamine contains three amino groups (reactive sites) compared to only two in urea, resulting in a higher degree of crosslinking in the final condensation products. The addition of formaldehyde to melamine is easier and more

Exhibit 3.n

Melamine-Formaldehyde Adhesives: General Characterization

Type	Synthetic thermosetting resin
Physical form	Film, liquid or powder form. Powder is mixed with water or water-alcohol mixture and sometimes supplied with a filler powder.
Shelf life (@20°C)	Film: 3 mth Powder: up to 2 yr
Working life (@20°C)	Film: 3 mth Powder mixture: up to 24 hr
Closed assembly time (@20°C)	Film: 3 mth Powder mixture: up to 24 hr
Setting process	Condensation polymerization with elimination of water
Process conditions	Application dependent, and within the range 10 hr at 50°C to 2 min at +90°C. Bonding pressures from 700-1000 kPa.
Coverage	General purpose bonding: 0.15-2 kg/m^2 plywood fabrication: 0.075-0.125 kg/m^2
Permanence	Melamine resins have superior resistance to water and better temperature stability than the related urea resins. Hot pressed melamine resins are especially resistant to boiling water, and are generally equivalent to hot pressed phenolics and resorcinol resins in water resistance and durability; low temperature, acidcatalysed, melamines are less durable than the phenolic resins. Resistance to biodeterioration is good. Melamine resins are comparable with many other materials and are used as fortifiers for urea-formaldehyde resins to promote their resistance to boiling water. Melamine-urea adhesives are intermediate between melamine and urea adhesives with durability proportional to melamine content.
Applications	Non-structural woodworking assemblies for exterior use. Scarf joint manufacture for casein bonded wood laminates. Fabrication plywood which is resistant to boiling water; melamine is employed as a urea-formaldehyde fortifier.
Comments	Bond strengths exceed the strength of the wood for very thin glue-lines. Film types display poor gap filling properties; powders benefit from the use of extender additives. Adhesives are light colored, non-staining and suitable for light woods and veneering. High temperature curing resins have long assembly periods and retain their low viscosity better than the low temperature curing types which develop tack quickly. For structural bonding, it is advisable to employ less expensive adhesives with more easily controlled curing processes. Bonded assemblies are usually conditioned for 1 day before handling.

Source: Shields, J. Adhesives Handbook, 1984.

complete than with urea. Each amino group in melamine
easily accepts up to two molecules of formaldehyde,
resulting in up to six molecules of formaldehyde
attached to each melamine molecule.

The manufacture of MF resins is carried out in the same
way as for UF resin production, involving the
condensation reaction of melamine and formaldehyde in
the presence of heat and an alkaline catalyst (usually
sodium hydroxide), resulting in a progressive
polymerization process that is halted while the resin
polymers are still water-soluble (See Exhibit 3.1).
Typically, the resin solution is then spray-dried, and
final curing is accomplished with the addition of heat.
Basic temperature and pH variables are similar to those
for urea manufacture, but the melamine-to-formaldehyde
molar ratio is usually about 1:2.5, compared to a urea-
to-formaldehyde ratio of about 1:1.5.

Melamine resins are most commonly supplied as dry,
stable powders without any catalysts or fillers.
Melamine-urea mixtures may contain fillers and
catalysts, if necessary.[125, 3]

In the wood adhesive area, MF resins are commonly used
in combination with UF resins when additional durability and bonding strength is required. Such products
include exterior-grade plywood, marine-grade plywood or
laminated timbers, and some particleboards and molded
wood waste products. The added moisture resistance of
a MF/UF mixture is greater than the proportional share
of MF resin; for example, a mixture with 50 percent MF
resin yields a bond with 90 percent of the moisture
resistance of a 100 percent MF bond. MF/UF mixtures
are produced either by blending dry UF and MF resin
powders, or by blending the two separate resins in
liquid solution and then spray-drying the mixture.
Melamine and urea can also be copolymerized with
formadehyde in the initial resin manufacturing stage.
The most common combinations contain 40-50 percent MF
by weight. Catalyst requirements to complete curing of
the UF component depends on the MF/UF proportions.
When the MF content is equal to or greater than the UF
content, no catalyst is used.

Market

MF resins are used in several adhesive and non-adhesive applications. A recent estimate of relative MF consumption included the following categories:[16]

- laminating resins   26%

- surface coating resins   24%

- molding compounds   18%

- paper treatment resins   12%

- textile treatment resins   8%

- adhesives and bonding
  (including insulation
    binding)   6%

- miscellaneous uses   6%.

In non-wood uses, MF resins modified with alcohols are used as cross-linking agents in the formation of industrial enamel coatings. Plastic dinnerware is the predominant end use for MF molding compounds. In paper products, MF resins are used to improve wet and dry strength and stiffness, and as crosslinking agents in binder pigment coatings.

Because MF resins form a colorless glueline, they are important in some architectural wood laminates and hardwood plywood applications. Although MF resins are much more water-resistant than UF resins, MF and MF-UF mixtures are still not considered as water-resistant as phenol- or resorcinol-formaldehyde resins.

The most important use of MF resins is in the "decorative laminates" sector. Formica is the best-known product in this sector. This market makes use of both the adhesive and non-adhesive properties of MF resins. MF resins are used because of their unusual chemical inertness, non-porosity, and nonabsorbency, which makes them resistant to most common liquids and staining agents. They are also resistant to scratching and are easy to color and mold.[125]

In 1983, the U.S. produced 180 million lb (dry weight) of MF resins. Major producers of these resins include

American Cyanamid and Reichhold Chemicals. Exports in 1983 exceeded 16 million lb, with leading markets in the Netherlands and Canada. Imports in this year were just under 2 million lb, with Canada, Israel, and West Germany as major sources.

Wood Adhesive Application Processes

By far the largest wood adhesive-related use of MF resins is in the manufacture of laminated paper products. Laminated products are made by impregnating paper or other fillers with a solution of resin. Multiple layers of impregnated sheets are cured in a press at high temperatures and pressures. Both decorative and industrial laminates are produced, although decorative laminates dominate. For decorative laminates, the body of the laminate consists of several plies of phenolic resin-impregnated kraft paper. This is overlaid with a MF resin-impregnated barrier sheet which is colored or on which a design is printed. This is covered with a clear overlay sheet containing MF resin.

In other wood products uses, resin application and manufacturing processes using MF resins are generally the same as hot-press UF applications, except that no catalysts are needed to cure MF resins. In hot-press processes where MF is used, temperatures are usually 240-260°F.[104, 125]

COMPARISON OF MAJOR CONVENTIONAL WOOD ADHESIVES

The four major wood panel adhesives have advantages and disadvantages which are traded off in the market place to determine market share. Exhibit 3.o summarizes some of the most important features of phenol-formaldehyde, resorcinol-formaldehyde, urea-formaldehyde, and melamine-formaldehyde resins. RF resins, the "ideal adhesives," form the strongest, most water-resistant bonds, are the most reactive, and cure the fastest at ambient temperatures. However, these resins are also the most expensive, limiting their market use to marine plywood and to mixtures with PF resins. In addition, RF resins are red-brown in color, preventing use in light-colored woods and veneers. They also and have a very short working life once hardeners are added. The RF resin market is not expected to grow much in the near term, although the possibility of increased use in metal and rubber bonding is currently being explored.

Exhibit 3.0

Comparison of Major Wood Panel Adhesive Resins

Conventional Adhesive Systems	Technical Attributes					Economic Attributes		
	Reactivity	Color	Bond Time	Cure Temperature	Bond Strength	Source of Supply	Cost ($/Lb)	Major End-use Application
Phenol-formaldehyde (PF)	Moderate	red	Intermediate	High (180°F)	Strong	Oil	.32-.37	- Plywood - Exterior particle-board
Urea-formaldehyde (UF)	Moderate	light	Slow	Ambient	Fair	Natural Gas	.16-.18	- Particle-board (Interior use)
Melamine-formaldehyde (MF)	Moderate	white	Slow	High	Good	Natural Gas	.58-.60	- Plastics - Particle-board
Resorcinol-formaldehyde (RF)	High	red-brown	Rapid	Ambient (70°F)	Very strong	Oil	1.65-1.80	- Marine plywood

Source: Adapted from Skeist, I. Handbook of Adhesives, 1977.

Phenol-formaldehyde resins are less expensive than RF resins and still offer a highly water-resistant bond. For this reason, PF resins capture the exterior-grade plywood market. PF resins do require heat to cure, and can also stain wood with their dark red color. The outlook for the U.S. phenolic resin market is for continued annual growth at 3-4 percent, tracking the projected construction industry growth.

Urea-formaldehyde resins are the cheapest of the four resins. However, UF resin bonds are susceptible to degradation with heat or moisture, so its use is limited to interior plywood and particleboard applications. Future environmental regulation of formaldehyde emissions could impede or eliminate such interior uses within the next 5-10 years. Therefore, future growth in the UF resin market is expected to be limited.

Melamine-formaldehyde resins are more expensive than UF and PF resins, but less expensive than RF resins. MF resin bonds have greater water resistance than UF bonds, although they are not as strong as PF and RF bonds. The main uses for MF resins are in decorative laminates, where their white color is desired, and in combination with UF resins to improve water-resistant qualities. The melamine resin market is expected to grow rapidly (7-12 percent annually) over the next few years, but this growth will result more from its use in moldings than in wood products.

MARKET STRUCTURE

The production and distribution system for synthetic wood adhesive resins in the United States seems to be sharply differentiated according to the size of the manufacturing operation and the type of product. In general, producer competition in the wood adhesive market is segmented between so-called bulk commodity adhesive uses and smaller, more specialized adhesive applications. The "commodity" market encompasses the relatively large-scale manufacture of plywood, particleboard, and related products. The smaller, more specialized markets include end uses such as decorative and structural laminating, wood molding, impregnated papers, and joint gluing.

Recent trends indicate that the production and distribution of commodity wood adhesives is becoming

increasingly concentrated, while the rest of the market
contains a number of active small resin producers and
adhesive formulators. In the commodity markets, there
are few if any intermediaries between the company that
produces the resin and the facility that manufactures
the final wood product. In such cases, the basic resin
product (which may already contain some non-resin
additives) is shipped to the plant, where a very site-
specific adhesive mixture is formulated. Representatives of the resin producer are often closely involved
in providing technical service and advice to the plant-
level personnel.

In the non-commodity wood product markets, the market
structure is more complex. There are a larger number
of producers, many of whom concentrate on only a few
specialized markets and/or resin products. There are
also a substantial number of intermediary adhesive
formulators, who modify the basic resin products and
produce a variety of specific adhesive mixtures that
are then sold to end users.

In recent years, the commodity wood adhesive resins
market (primarily PF and UF resins) has come to be
dominated by 2-3 major resin producers. Borden
Chemical and Georgia Pacific are clearly the market
leaders. Chembond Corporation is also active in the
market, but the relative scale of its output is
uncertain. Within the last four years, at least three
producers -- Gulf, Reichhold Chemicals, Pacific Resins
and Chemicals, and Monsanto -- have left the commodity
market. Monsanto's PF business was sold to Borden in
1983 and Georgia Pacific acquired Pacific Resins and
Chemicals in 1981.

The trend toward concentration in the commodity market
is driven by competitive pressures in a mature and
cyclical market and significant competitive advantages
among the remaining companies. Most of the companies
who left the commodity adhesive market cited a
combination of weak prices and high production costs.
Another factor seems to have been the need to maintain
high levels of end-user technical service to remain
competitive in the market.

On the other hand, the companies still in the market
have presumably been able to build company-specific
competitive advantages. Borden has strong backward
integration into most of the main chemical raw

materials needed for resin production, based on its integrated "petro-condo" complex at Geismar, Louisiana. Georgia Pacific, with about 17 percent of the U.S. plywood and particleboard market, used a large share of its adhesives output in its own manufacturing operations. Until 1984, Georgia Pacific also had a large chemical division. (GP's chemical operations, with the exception of the resin products division, were sold to Georgia Gulf). Chembond's situation is less clear, but it is the U.S. subsidiary of a Finnish wood adhesives manufacturer (Priha Oy).[119]

Exhibit A.2 in Appendix A presents additional detail on the various U.S. thermosetting adhesive producers.

4 Nonconventional Wood Adhesive R&D

The conventional formaldehyde binders for composite
wood panel products which were described in Chapter 3
have several shortcomings that have become increasingly
important over the last decade. For example, phenol-
formaldehyde forms a brittle glueline that for some end-
use applications can become detrimental over time.
Urea-formaldehyde resins are coming under increasing
attack because of formaldehyde emissions and resulting
indoor air pollution problems. Formaldehyde binders
also have long pressing times, intensive energy
consumption, and health risks for workers. Attempts to
formulate new non-conventional bonding systems have
explored ways to emulate the behavior of conventional
adhesives. There is some concern over the feasibility
of this approach, particularly when problems already
exist with current resins. Therefore, some industry
representatives suggest that the goal of non-
conventional wood adhesive research should be to
develop completely new bonding systems with enhanced
characteristics.

In the U.S. and many other countries intensive efforts
have been made to develop new binders that would
overcome these problems. Most of this research has
concentrated on three potential derivatives for wood
adhesives: lignins, tannins and isocyanates. Wood
surface activation for bonding without adhesives and
adhesives derived from other types of natural substances like foliage have also been tested, but these
three areas -- lignins, tannins, and isocyanates
-- have shown the most promise as potential substitutes
for conventional adhesives. Also, they have received
the greater share of attention and funding. Certain of
these non-conventional wood adhesives have already been
commercially used in wood panel plants in Europe, South
Africa and Australia. Specifically, lignin-based wood
adhesives are commercially produced in Finland, tannin-
based adhesives are used in South Africa for wood panel
products, and isocyanate-bound composite boards are
manufactured in West Germany.

Descriptions of these three major areas of R&D in non-conventional bonding systems are presented below.

LIGNINS

Lignin is the non-carbohydrate portion of extractive-free wood, accounting for 20-30% of the weight of wood (see Exhibit 4.a).[62] About 70% of this lignin exists in the intercellular regions of the wood, and the remainder is located in the cell wall, dispersed among groups of carbohydrates. In the manufacture of highly purified pulps for papermaking, it is necessary to remove all of the lignin with minimum loss or degradation to the carbohydrate cell wall.

Pulping, a process whereby the individual fibers of wood are liberated through the application of extreme heat or pressure, causes a fundamental alteration in the structure of lignin. Historically, the lignin fraction produced during pulping has either been considered a waste product, an energy source, or most recently, a potential feedstock for aromatic-based chemicals. The complexity of the lignin polymer has restricted more effective utilization of this resource.

Technical lignin (i.e., natural lignin modified through industrial processing) is available from several chemical processes. In the United States, the majority of commercial lignin is produced from the kraft pulping process. Kraft pulping is the most widely used pulping process in the United States because it is adaptable to almost any wood species and because pulp from the kraft process can be easily bleached.

Lignosulfonates are commercially available in the U.S. market on a much smaller scale. While lignosulfonates can be formed from kraft lignin by sulfonation, they are a direct by-product of sulfite pulping, which is used only on a limited basis in the United States. Lignosulfonates are extensively used overseas, where air pollution control standards restrict the use of kraft pulping and sulfite pulping is more popular. Other sources of technical lignin are organosolv, steam explosion, autohydrolysis, rapid steam hydrolysis, wet oxidation, and biomass conversion.   (See Exhibit 4.b

Exhibit 4.a

Chemical Composition of a Sample of North American Trees
(percentage)

	Hardwood			Softwood		
	Aspen	White Birch	Beech	Balsam Fir	Jack Pine	Eastern Hemlock
Lignin*	20.9	19.0	22.1	30.0	28.6	32.5
Cellulose*	42.7	43.1	44.0	42.2	42.1	41.6
Ash*	0.4	0.2	0.4	0.3	0.2	0.2
Other Components	36.0	37.7	33.6	27.5	29.1	25.7

*Values based on oven-dry extractive-free wood.

Source: Kent, James A. Riegel's Handbook of Industrial
Chemistry, 1983.

Exhibit 4.b

Summary of Technical Lignin Sources

Source of Lignin from Industrial Processes	Availability of Process	End-Product
Sulfite Pulping	Commercial - limited	Pulp
Kraft Pulping	Commercial - extensive	Pulp
Organosolv	Not Commercial	Pulp
Steam Explosion	Commercial - limited	Hardwood
Autohydrolysis	Not Commercial	Pulp
Rapid Steam Hydrolysis	Not Commercial	Pulp
Wet Oxidation	Commercial - limited	Condensed Wastes
Biomass Conversion	Commercial - limited	Fuel

Source: Data compiled from extensive industry
           interviews by Hagler, Bailly & Company

for the level of commercialization and uses for these processes.)

Chemical Composition

Lignin is composed of phenylpropane units that are linked through carbon-to-carbon or carbon-to-oxygen (ether) bonds. It is not a specific chemical entity, but rather a complex, heterogeneous polymer that describes a class of related materials. Softwood and hardwood lignins have considerable differences in their molecular structure. As a result, lignins are typically referred to by their plant source, such as hardwood or softwood, or by method of isolation, such as kraft lignin or lignosulfonates.

The objective of chemical pulping is to solubilize and remove the lignin portion of wood. The majority of pulping processes in use today are either alkaline (kraft) or acidic (sulfite). In general, pulping is a cooking process occurring in batch digesters. Conditions for pulping vary according to wood species and the quality of pulp to be produced.

Naturally occurring lignin is neither hygroscopic (the ability to absorb or attract moisture from the air) nor soluble in water. Under acidic conditions, lignin will react with bisulfite ions to form lignosulfonates, which are soluble in water and can easily be separated from the pulp.

Sulfite pulping is a simple process that uses inexpensive chemicals in fairly limited amounts. Its major drawback is that it does not produce as strong a pulp product as the kraft process. No chemical recovery is necessary in sulfite pulping. Nonetheless, many sulfite mills have been replaced by sulfate mills because of high effluent loads with no ready means of chemical recovery. New technology is emerging for the economic recovery of chemicals from sulfite pulping on a scale similiar to that of sulfate mills. Consequently, some sulfite mills are being restored.[29]

Technical spent sulfite liquor contains 50-60% lignosulfonates, about 30% carbohydrates, and up to 20% inorganic salts depending on the pulping conditions.[62]

In acid sulfite pulping, sulfonation and hydrolysis together result in lignin solubilization. Consequently the lignin becomes an integral part of the pulping liquors. The lignin which is recovered from spent sulfite liquors is a sulfonate. Little methoxyl group loss takes place, and the sulfonate groups, which predominate in the benzylic position, impart a strong hydrophilic character to the lignin.[91]

The kraft pulping process uses a mixture of sodium hydroxide and sodium sulfide as the active chemical agents. It produces the strongest pulp commercially available. In kraft pulping, the high alkali charge requires that the chemicals be recovered and re-used since the chemical costs associated with the process are significant. The yield of pulp is only about 45% of the original wood weight, and the remaining organic residues must be eliminated. The black liquor separated from the pulp is concentrated in evaporators to 60-65% solids. At this concentration, the quantity of dissolved organic compounds from the wood (lignin and carbohydrates) is sufficient to allow it to be burned in the recovery furnace. Since the chemical system is a closed one, costs and pollution are minimized.[29]

Burning of the black liquor in kraft pulping does emit hydrogen sulfide and mercaptans that can create serious air pollution problems. In recent years, techniques, such as black liquor oxidation have been put in place to improve emissions control.

In kraft pulping, the combination of strong base conditions and sulfides tends to break down some of the ether linkages and increases the loss of reactive methoxyl groups. Molecular fragmentation of the lignin polymer is more severe in kraft pulping than in sulfite where the formation of phenolates results in solubilization. The kraft process, however, does produce a lower molecular weight lignin than the sulfite process.

In both pulping processes, oxidation and condensation of reactive groups can occur. Because of the variety of chemical environments used in pulping, the qualities of technical lignin vary accordingly. Changes in the molecular structure of lignin caused by pulping affect the reactivity, or the availability of reactive sites in various technical lignin.

Lignin is a thermoplastic material whose softening point is influenced by exposure to water. Chemical processes, such as oxidation, condensation, and hydrolysis alter the functional groups comprising the lignin polymer. As a result, the thermal properties of lignin are affected during delignification. The ability of lignin to flow and penetrate into wood at a specific rate in relation to temperature, time, and pressure will influence its utility as an adhesive.

In the future, technical lignins may also be available from an organosolv or other processes. Under the milder conditions of the organosolv process, condensation during pulping among the lignin subgroups will be reduced. However, the resulting product will have fewer hydroxyl sites available. This will retard further polymerization of these materials into cross-linked adhesives. Technical lignins from steam explosion are thought to be less reactive than kraft or sulfite lignins.

Adhesive Utilization

Many efforts have been made to utilize lignins for a variety of end-products, but they have met with limited successs. The earliest application of lignin as an adhesive was the pressing of decayed wood to form a product of high strength and limited water resistance. Decay promoted by fungal attack had physically and chemically restructured the lignin into an active adhesive.

Chemically, the curing of lignin requires cross-linking. When spent sulfite lignins are used to form an adhesive, they must be converted to an insoluble state during the curing process. Crosslinking reactions decrease the solubility and swelling capacity of lignin. The majority of patents available in recent years achieve cross-linking through condensation reactions. To achieve condensation, lignin generally is exposed to either high temperatures and long heating times or mineral acids which, when applied, can cause structural changes and sometimes charring of the wood particles.

Much research has focused on lignosulfonates for adhesives because they are available at a low cost, being a waste product. The major drawback in using lignosulfonates is hydroscopicity resulting from the presence of sulfonic groups. For practical applications, this has translated into longer pressing times in order to form water proof bonds (autoclaving is also required). Scientific attention was directed to this problem in the 1960s when lignosulfonates were commercially introduced in Denmark to bond wood chips.[21]

In Canada, some improvements have been made in the pressing times of lignosulfonate-based adhesives by the addition of small amounts of acid-curing phenol-formaldehyde resin. However, it was found that high pressing temperatures were still necessary.

Similar effects could be accomplished through electrodialysis treatment. Electrodialysis was used to reduce cations to produce a sulfite lignin adhesive with a pH of less than 1.0. Although good bond quality was achieved through this method with acidified ligno-sulfonates in waferboard, the long-term effects of wood degradation countered any improvements in bond strength. With reduced acid content, bond stability was greater, but cure time was markedly increased. An attempt was also made to use lignosulfonates at a pH of 3.0 in particleboard. While the dimensional stability of this adhesive was feasible, plant operation of this Pederson process proved that its long press times and post-treatment were uneconomical. However, the performance level of the adhesive was considerable for a non-synthetic resin bonding system.

Recent work with ammonium lignosulfonate suggests that the reducing sugar level present in lignosulfonate adhesives influences bond quality. The pressing conditions of high temperature and long press times have probably converted some of these sugars to furfuryl derivatives which during polymerization create a highly thermal and hydrolysis-stable polymer.

Another approach to using lignosulfonates has been to separate the high molecular weight species by ultra-filtration and use these species as co-reactants with a low molecular weight PF resin. The resulting resin system contains about 50% lignosulfonate. Cure times have been significantly shorter.

One approach to overcoming lignins' tendency for longer
cure times and higher cure temperatures involves cross-
linking between lignin units by oxidative coupling.
Oxidants such as hydrogen peroxide catalyzed by sulfur
dioxide, ammonium chloride, or potassium ferricyanide
cause a strong exothermic reaction with lignosulfonates. Mainly, this adhesive system has been applied
to particleboard in pilot plant runs at the University
of California/Berkeley.

Kraft lignins have a high inorganic content.
Acidification and removal of these ions can produce a
modified lignin product which has considerable adhesive
qualities. The key to kraft lignin utilization is
purification, because the presence of inorganic ions
reduces the water resistance of the bond. Along with
purification, addition or condensation reactions are
also required to enhance bond capabilities. Even with
these modifications, the complex structure of lignin
and its low reactivity have deterred adequate
formulation of highly crosslinked adhesives.

In the United States and Canada, attempts have been
made to co-react lignin and formaldehyde. These
approaches paralleled similiar efforts in Japan at the
Hokkaido Forest Products Laboratory by Abe. The
resulting adhesive systems showed some improved
adhesive qualities, but still required the addition of
30% or more of a low-molecular weight PF resin to
achieve durable bonds in waferboard.

The Karatex process developed at the Finnish Pulp and
Paper Institute in Helsinki isolates a high molecular
weight fraction from kraft liquor using ultrafiltration. The lignin can then be substituted for almost
70% of the phenol in conventional PF adhesives. The
adhesive was tested on particleboard, plywood, and
fiberboard. While the Karatex system has been employed
with some success in Europe, intolerance of long
assembly times in the domestic forest products industry
has stood as an obstacle to its commercialization in
the United States.

Another approach to lignin utilization involved the
methylolation of kraft black liquor at high alkalinity
followed by acidification and precipitation. The
dispersed precipitate was able to replace up to 70% of
the PF resin in an acid-curing adhesive system.

On the whole, most attempts to use lignin to produce self-condensing durable adhesives have been unsuccessful. Under the most ideal conditions thus far achieved, only 50-70% of a PF resin can be replaced by lignin. This is not surprising considering the extremes of temperature, pressure, and pH that lignins are exposed to during pulping. Pulping has the effect of moderating reactivity and increasing structural complexity to the point of limiting the use of lignin as an adhesive.

To arrive at a lignin material directly substitutable for phenol, much more basic research work is required. An alternative approach to direct substitution may be to break down the lignin polymer to its lower molecular weight constituents for formulation into phenol. By breaking down the lignin polymer through oxidative alkaline hydrolysis, hydrocracking, hydrogenolysis, demethoxylation, and other solvent recovery schemes, structural complexity and variability are reduced and reactivity increased. With regard to hydrocracking, it may be feasible to link a hydrocracker reactor to the pulping process and produce phenol from the cresols and benzene obtained from the hydrotreated lignin.[43]

Other alternatives include hydrogenolysis and demethoxylation which leads to 4-methyl-catechol, a compound very reactive with formaldehyde. In Japan at Mie University, Funako and Abe developed a process whereby lignin is reacted in the presence of boron trifluoride. The yields of catechol obtained from this process were quite high. Catechol provides a useful starting material for phenolic-type resins.

For commercial applications, it is important that any process manufacture a simple and chemically clean product with high yields. Modified lignin as a direct substitute may never satisfy industry's need for a consistent product. Also, lignin research has shown that most lignin-based adhesive systems require longer press times or higher cure times regardless of the extent of acid or base catalysis. Again, extensive interviews with industry representatives indicate that increased press times will be unacceptable.

TANNINS

Tannin refers to a widely occurring group of substances derived from plants. In their most common usage, these substances are capable of rendering raw hides into leather. Common tannin or tannic acid is found in a variety of nuts, tea, sumac, and bark. It is a complex, dark polyhydroxy phenolic compound.

As a bark product, tannins are readily available. At mill sites where bark is a waste product of sawmilling and other panel making processes, large quantities of bark are presently used as an energy source, in land-fills, or for horticultural purposes. On average, bark contains 15-30% polyphenols, 20-30% lignin, 30-45% carbohydrates and 1-3% fats and waxes.[91] These values are dependent upon wood species, location, age, and exposure.

Since two-thirds of bark is composed of aromatic polymers, the natural availability of such substantial amounts of low-cost polyphenols has spurred significant R&D activities over the years. This research has concentrated on formulating commercially acceptable chemical products from tannins. Research activities in this area over the last 40 years have taken place in Australia, New Zealand, South Africa, and North America. Limited commercial success has been attained in Australia and South Africa with wattle and radiata pine species.

Chemical Composition

The major class of polyphenols in bark are condensed tannins. Tannins can comprise anywhere from 3-30% of bark depending on the factors listed above. When tannin is extracted in an aqueous solution, it often is closely associated with carbohydrate groups. The basic structure of condensed tannin is displayed in Exhibit 4.c.

Bark behaves quite differently from wood because of its unique anatomy, large aromatic composition, and low fiber content. The presence of living and dead tissue leads to extremely variable water contents and specific gravity ranges among and within species. The inherent variability of bark species is reflected in the

Exhibit 4.c

Basic Chemical Structure of Tannins

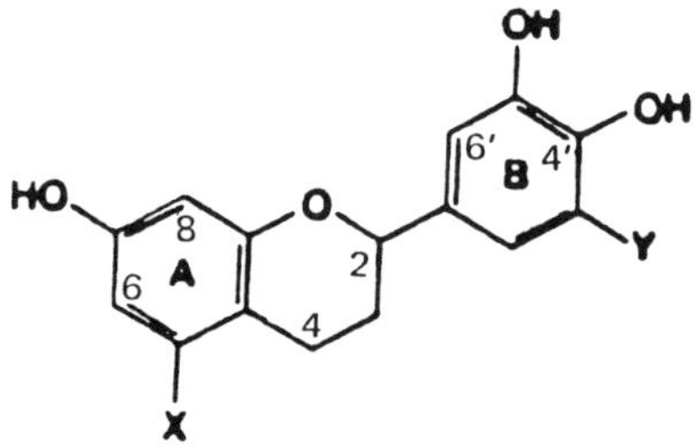

Source: Oliver, John F. <u>Adhesion in Cellulosic and Wood-based Composites</u>, 1981.

composition and yields of bark extracts, particularly tannins.[91]

Tannin extract solutions are colloidal. Dramatic variations in particle size distributions occur with changes in concentration, temperature, pH, and added salts, with larger particle sizes tending to be hydrophobic. The strong molecular association present in tannin solutions has resulted in some ambiguities in molecular weight values.[91]

Adhesive Utilization

Research on bark utilization has involved bark particles, bark powders, and tannin extracts. It has been found that bark particles pressed at below 355°C will form boards that are water-resistant without the application of an adhesive. At higher temperatures, polymerization of phenolic extracts and lignin within the bark produces a higher quality exterior-type board. Dr. S. Chow of Canadian Forest Products Ltd. found that the strength, dimensional stability, and moisture adsorption of these boards were similar to or better than bark boards made with 4 or 7% UF or PF resins, with pressing schedules similar to commercial particleboard production.[21]

Bark powders have been used as the hydroxy group source in the formation of epoxy or polyurethane systems or as partial replacement for UF or PF resins in adhesive mixes. In the latter cases, viscosity control was a limiting factor.

Interest in tannin extracts was prompted by the work of Dalton in Australia and Roux in South Africa. Roux researched the fundamental chemical characterization of wattle bark tannin. Dalton took wattle extracts and reacted them with formaldehyde to create a water-resistant wood bond after hot-pressing. Despite promising results, boiling water testing proved that the bond was not sturdy enough for exterior use. Also, viscosity control was poor. Further development by Plomley illustrated that wattle tannin adhesives could be used in bonding particleboard and plywood; however, in plywood applications Plomley had to fortify the adhesive with a PF or PF-RF resin to achieve greater exterior durability.

Tannins from quebracho, mimosa, mangrove, and radiata
pine have also been used to bond lignocellulosic
materials. Recent findings suggest that tannin extract
purity has a significant influence on bond strength and
water-resistance. As little as 20% carbohydrate
material in the extract reduced bond strength by almost
half. For this reason, reduction of these carbohydrate
units by fractionation, precipitation, or use of
organic extraction solvents has become increasingly
prevalent.[91]

Compared with wattle tannins, bark extracts from
spruce, hemlock, and southern pine species have the
disadvantage of lower yields of polyphenols. The
tendency with this group of tannins is to have too slow
of a reactivity with formaldehyde to offer good
adhesive properties. Attempts to lengthen pot-life
have been achieved by using paraformaldehyde or
hemamethylenetramine instead of formaldehyde solutions.
Steiner and Chow showed through thermal analysis and
plywood-bond tests that higher temperatures during
pressing improve western hemlock tannin-formaldehyde
bonds by providing sufficient energy to extend cross-
linking. It is believed that higher temperatures
enhance reactivity in the polyflavonoid ring.[91]

Extensive work by Roux, Plomley, Hillis, and Hemingway
has shown that, through understanding bark chemistry
and the reactivity of model compounds, improvements can
be made in the handling and polymerization of bark
tannins. From the experience of South Africa, it
appears that standardized procedures for forestation,
growth climate, harvesting, extraction, and
concentration techniques are necessary to increase
tannin reactivity and reduce glueline brittleness.[91]

Modifications of tannin extracts have involved the
grafting of phenol monomers onto the polyflavonoid and
then co-reacting with formaldehyde in situ to produce a
tannin-resol system or the fortifying of tannin extract
with a PF or UF resin. Both approaches produce
improved adhesive properties with reduced press times.
The success of these modifications relates to the
increase in reactive sites for formaldehyde and the
ability of these low-molecular weight additives to
bridge the hindered reactive sites of the larger tannin
molecules. It is still uncertain whether the fortifying resin co-reacts with the polyflavonoid or only

forms an intertwining cured matrix with the bark
polyphenols. Hemingway concluded that the effective
use of PF fortifiers requires: a high methylol content
PF of low molecular weight, a solution pH less than 8,
methylol phenols partially condensed with tannin prior
to pressing, and in particular no free formaldehyde
present during the early stages of adhesive
formulation.

Although advances in tannin-extract utilization have
been promising during the last 25 years, product
variability between trees and species, with different
inherent reactivities, are factors which will always be
present. For these reasons, conventional techniques
will likely require a fortified-tannin adhesive to
achieve a high quality, durable bond in the foreseeable
future. However, additional knowledge of tannin
structure and its effect on reactivity, together with
higher pressing temperatures and more effective
catalysts, may afford a tannin-only adhesive for
bonding lignocellulose products of intermediate
durability.[91]

ISOCYANATES

Isocyanates have been used in the manufacture of
polyurethane resins for nearly 30 years; however, their
use in wood adhesives is relatively new. Application
of isocyanate as a binder for particleboard was
explored as early as 1951 by Deppe and Ernst. As a
result of their work, the first mill producing
particleboard completely bonded by isocyanate began
operation in the mid-seventies in West Germany.

Although isocyanate differs from other non-conventional
adhesives because it is petroleum-derived, the novelty
of its adhesion system and its improved technical
performance increase its potential as an important
substitute for thermosetting resin adhesives in the
near term.

All isocyanates of industrial importance contain two or
more isocyanate groups per molecule to enable polymers
to be formed by condensation with other polyfunctional
molecules. Only a limited number of isocyanate
compounds have gained technical significance. Their
selection is governed by such factors as reaction
behavior, physiological properties, availability of

starting materials, and economies of production. The preferred isocyanates for the composite wood panel industry are those based on diisocyanate diphenyl-methane (MDI). Isocyanates, as oil-based products, are derived from naphtha as shown in Exhibit 4.d.

Chemical Composition

The chemistry of isocyanate and its ability to bond to wood are inextricably tied to the formulation of poly-urethanes. The two major reactions in the preparation of polyurethane are the reaction between isocyanate and hydroxyl-containing compounds and that of isocyanate and water. Other reactions that foster wood adhesion are the reaction of the isocyanate group with amines and the formation of linkages with urethane. These reactions cause branching and crosslinking. Cross-linking also occurs from trimerization, which leads to the formation of isocyanurate rings that are hydroly-tically and thermally very stable. Isocyanates form various length bridges with the alkyl and phenolic hydroxy groups in wood cellulose and lignin. The reactions that take place between isocyanate and wood are displayed in Exhibit 4.e.

Through the formation of polyureas, variation in the isocyanate polymer length occurs and contributes to successful wood bonding. One impediment to this process is water, which must be carefully controlled. If too much water is present, a large amount of isocyanate is deactivated and the system loses many of its adhesive qualities. The urethane linkages become too few in number to create adequate bonding between the wood surface and the adhesive.

In most cases, isocyanates require a catalyst to increase the reaction rate and to establish a proper balance between the wood-to-adhesive crosslinking (hydroxyl-isocyanate reaction) action and polyurea formation (isocyanate-water reaction). Also, by increasing the reaction rate, catalysts induce a more rapid completion of the reaction, thus creating a more adequate glueline. Proper curing of this type results in optimal adhesive qualities -- greater strength and weatherability.

Exhibit 4.d

Manufacturing Route to Isocyanate

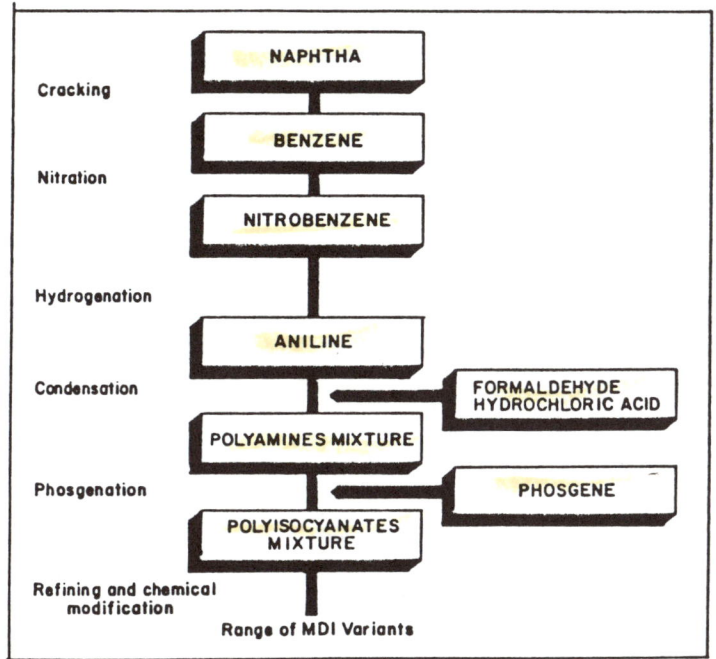

Source: Ball,   G.W.   International   Symposium   on
Particleboard, Washington State Unversity, 1981.

Exhibit 4.e

Reactions Occurring During Wood Bonding With Isocyanate

1. Polyurea Formation:

a)

$$OCN - \langle \rangle - CH_2 - \langle \rangle - NCO + 2 H_2O \xrightarrow[slow]{} H_2N - \langle \rangle - CH_2 - \langle \rangle - NH_2 + CO_2$$

b)

$$OCN - \langle \rangle - CH_2 - \langle \rangle - NCO$$

$$+$$

$$H_2N - \langle \rangle - CH_2 - \langle \rangle - NH_2 \xrightarrow[fast]{} \sim NHCONH - \langle \rangle - CH_2 - \langle \rangle - NHCONH - \langle \rangle - CH_2 - \langle \rangle - \sim$$

Polyurea

2. Reaction with Cellulosic Hydroxyls:

$$\rangle\!\!\!\rangle - OH + OCN - \langle \rangle - CH_2 - \langle \rangle - NCO \longrightarrow \rangle\!\!\!\rangle - OCONH - \langle \rangle - CH_2 - \langle \rangle - \sim$$

Wood   Urethane

Source: Ball, G.W. International Symposium on Particleboard, Washington State University, 1981.

Adhesive Utilization

A long standing disadvantage in the use of isocyanate
for wood adhesion was its tackiness and characteristic
of sticking to the pressing equipment. For this
reason, the development of release agents to prevent
costly buildup of the adhesive on the equipment has
been as important a factor in the commercialization of
isocyanate for wood bonding as formulation of the
adhesive itself.

Further problems with isocyanates were caused by its
high viscosity. The efficient distribution of the
adhesive was impeded by its viscosity, particularly
since it requires very low levels of adhesive application. Deppe produced improved boards by reducing the
viscosity of isocyanate with solvents or by applying
additional heat. This proved to be uneconomical from a
production viewpoint. Recent developments by
isocyanate manufacturers, however, have overcome this
distribution problem, leaving its high viscosity a cost
advantage since only very low adhesive levels are
required. (See Exhibit 4.f for a list of domestic
manufacturers of isocyanate.) Other recent technical
improvements include:

● water emulsifiable MDI isocyanates

● release agents for spraying the surfaces of the
  caul plates

● self-releasing isocyanates.

Isocyanate was recognized for its adhesive qualities as
early as 1932 when Hartmann modified cellulose by
reaction with isocyanate. Beginning in 1951, Deppe's
work illustrated that isocyanate has excellant moisture
resistance in wood particleboards. Unfortunately,
recent results from Washington State University in
conjunction with the Forest Products Laboratory in
Madison, WI revealed that when the wood moisture
content was greater than 12%, the bond strength of
waferboard declined considerably. On the positive
side, isocyanates have been shown to be superior to
conventional adhesives in that they require no buffers
or hardeners, and introduce neither acid or alkali to
the board which can lead to degradation. Also,
isocyanates are not formulated with formaldehyde, nor

Exhibit 4.f

U.S. Isocyanate Manufacturers

- Arco Chemical

- BASF Wyandotte

- Mobay Chemical

- Rubicon Chemical

- Upjohn

Source: Wilson, James B. International Symposium on Particleboard, Washington   State   University,   1981.

do they possess materials from which formaldehyde can be generated at a later stage.

Wittmann has shown that the important reactions occurring during the pressing of particleboard using isocyanates binders are the reactions with water to form hydrolysis-resistant polyurea, a polymeric glue, and with the hydroxy groups present in wood to form covalent urethane links between the wood particles and the polyurea glue.

Other approaches to the modification of isocyanate include the work of Alan Lambuth at Boise Cascade and Chung Hse at the Southern Forest Experimental Station. Both men have tried to incorporate lignin and other natural or synthetic polymers into an isocyanate system to control its reactivity. One of the most interesting attempts at enhancing the performance of isocyanates has been conducted by Pogel, Luckman, and Galagher. They have developed aspects of the emulsion polymer isocyanate, a two-component adhesive system where the emulsion system encapsulates the reactive chemicals and the working life of the adhesive is prolonged. When the adhesive is applied to wood under pressure, the reactive chemical is freed to react with the moisture in the wood.

In North America, flakeboard, medium density fiberboard, and waferboard have the potential for commercial manufacture with isocyanate. Several pilot plant runs have demonstrated isocyanate's effectiveness in poplar waferboard. Isocyanate has been found to successfully bond waste crops, like bagasse, which is abundant in developing countries. Structural composite board is another product that bonds well with isocyanates. Elingson Timber Company of Baker, Oregon, has produced in limited quantities an isocyanate board commercially known as Elcoboard.

PLAYERS IN NON-CONVENTIONAL WOOD ADHESIVE R&D

To gain an understanding of the organizations that develop innovative processes for the formulation of wood adhesives, key R&D players were identified both in the international and domestic market. For those players that were accessible in France and the United Kingdom, person-to-person interviews were arranged by Hagler, Bailly staff members in-country at the time of

this project. For other overseas players, letters of
inquiry were sent out during the course of the project
and a literature search was conducted for technical
journals on adhesives and forestry in English, French,
and Russian. In the domestic market, initial identification of major players was obtained through several
directories, including the Encyclopedia of
Associations, the Adhesives Age Directory, and the
Thomas Register. Once the major R&D players were
identified for the United States, an extensive series
of interviews was completed. Location of key players
through a literature review of articles and advertise-
ments as well as referrals by interviewees extended the
number of contacts to well over 150 individuals.

Twenty-four organizations were found to be actively
engaged in non-conventional wood adhesives R&D in the
United States. Three other players have been identified as past actors in the field. (See Exhibit 4.g for
a list of domestic R&D players.)

The majority of these key players are universities,
whose research efforts are primarily supported by
monies from the Federal government (see Exhibit 4.h).
In some instances, state governments have also promoted
and funded this type of research, but they are minor
players in wood adhesives R&D. Two examples of state
sponsorship in wood adhesives R&D are the work of Dr.
Gary McGinnis at Mississippi State University and that
of Dr. P. E. Laks at Michigan Technical University.

Most of the individuals doing research on wood
adhesives at universities found it difficult to obtain
corporate funding. Their impression was that industry
was basically not interested in R&D in this area.
However, two university centers -- Washington State
University and the University of California/Berkeley
-- have developed ongoing working relationships with
corporate sponsors who support their work in non-
conventional bonding systems. In fact, both have
received corporate funding for pilot plant runs to test
the practical feasibility of their non-conventional
adhesive systems. Consequently, the individuals
interviewed at these institutions were in the minority
when they expressed their views that the level of
cooperation from industry was quite high. Washington
State has a broadly based program in wood adhesives,
having done work in lignins, tannins, and isocyanates.

Exhibit 4.g

Domestic R&D Players in Alternate Adhesives

R&D Players	Alternate Adhesive Research			
UNIVERSITIES	Lignin	Tannin	Isocyanate	Other
Auburn	X			
Colorado State	X			
Michigan State		X		
Mississippi State	X			
North Carolina State	X			
Oregon State		X		
Texas A&M				X
Univ. of Calif/ Berkeley	X			
Univ. of Idaho	X			
Univ. of Illinois	X			
Univ. of Washington	X	X		
Univ. of Wisconsin	X	X		
Virginia Poly- technical Inst.	X			
Washington State	X		X	

GOVERNMENT AGENCIES AND GOVERNMENT-SPONSORED LABORATORIES

Battelle – Pacific Northwest Laboratories	X			

Exhibit 4.g (continued)

Domestic R&D Players in Alternate Adhesives

	Lignin	Tannin	Isocyanate	Other
*National Science Foundation	X			
SERI	X			
Tennessee Valley Authority				X
USDA Forest Service:				
Forest Products Lab	X			
Southern Forest Experimental Sta.	X	X		
PRIVATE SECTOR FIRMS				
ARCO			X	
Boise Cascade	X		X	
+Hydrocarbon Research	X			
Mobay			X	
+Reed Lignin	X			
UpJohn Chemical			X	
Westvaco	X			
+Weyerhaeuser	X	X		

+ No longer active in designated area.
* Only a sponsor of R&D.

Source: Data compiled from extensive industry
          interviews by Hagler, Bailly & Company.

Exhibit 4.h

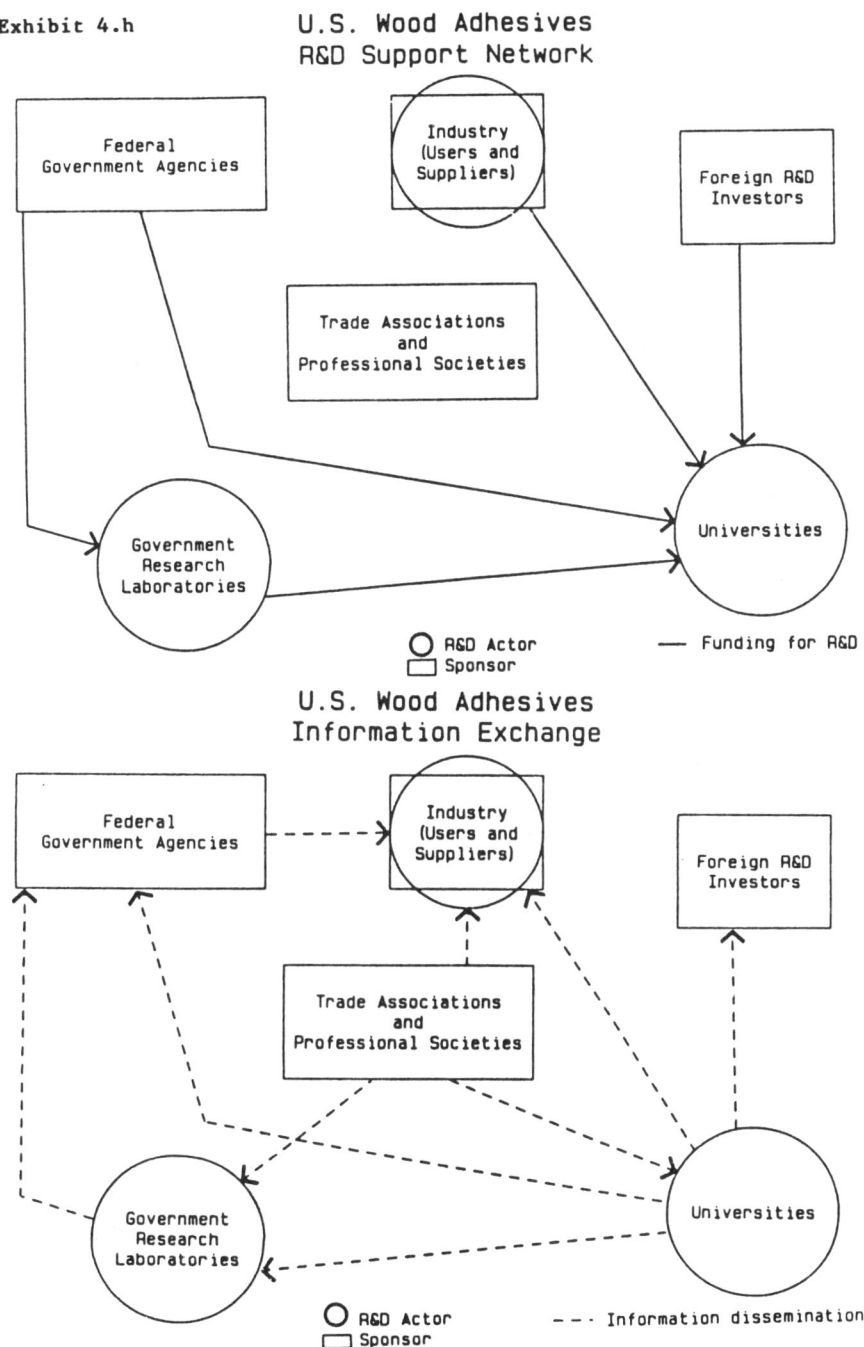

U.S. Wood Adhesives
R&D Support Network

Federal Government Agencies

Industry (Users and Suppliers)

Foreign R&D Investors

Trade Associations and Professional Societies

Government Research Laboratories

Universities

○ R&D Actor
▢ Sponsor

— Funding for R&D

U.S. Wood Adhesives
Information Exchange

Federal Government Agencies

Industry (Users and Suppliers)

Foreign R&D Investors

Trade Associations and Professional Societies

Government Research Laboratories

Universities

○ R&D Actor
▢ Sponsor

– – · Information dissemination

Source: Hagler, Bailly & Company

The University of California's program has focused on lignosulfonates and wood surface activation.

Government agencies are also active in wood adhesives R&D. Six separate government entities are doing R&D related to wood adhesives. The most active government organization in this area is the Department of Agriculture, whose Forest Service research laboratories have been active in wood adhesives R&D for over 50 years. While they actually conduct their own research, they also fund research through universities. In fact, they are one of the major sponsors of R&D in this area. In most cases, when government agencies fund research through universities, the university partner will do the fundamental chemical characterization of the substance, residue, or extract and the government agency will do the practical adhesive formulation and testing.

Some agencies strictly disperse research dollars to individuals with promising ideas. The National Science Foundation runs a biomass program from which it has funded lignin research at the University of Washington. Other government organizations have gotten involved in non-conventional wood bonding as a result of other processes they have developed. For example, the TVA has a wood-to-ethanol process from which lignins are a by-product. If these lignin residues could be used in a value-added chemical application, the process operation costs could be further justified.

Private sector actvity in R&D was the hardest to characterize because corporate representatives were somtimes unwilling to discuss their R&D activities. However, information on companies could usually be obtained from competitors or academic recipients of corporate funding. Most industrial R&D on non-conventional wood adhesives occurred in the late seventies during the oil crisis. Today, much of that activity has tapered off, particularly after the 1982 recession. In that time frame, many of the large, diversified wood producers sold their resin manufacturing units and became less inclined to do wood adhesives R&D. The manufacturers of conventional wood adhesives, while they may have been active in R&D during the oil crisis when their petroleum feedstock was unavailable, have little reason to pursue non-conventional adhesive R&D during an oil glut. It is suspected that a low

level of R&D is taking place in certain wood panel
manufacturing firms. These firms have expressed their
willingness to test new adhesives developed at several
universities. This has also been true for the large
wood adhesive manufacturers.

The greatest impetus for non-conventional adhesive R&D
may reside with non-traditional players. An excellent
example of this is the case of isocyanates.
Isocyanates are produced by the manufacturers of
polyurethanes. The producers of polyurethanes are
generally not the producers of phenol- or urea-
formaldehyde resins. In the case of wood adhesives,
chemical companies that were not the traditional
suppliers of wood adhesives were found to be actively
involved in non-conventional wood adhesives research,
attempting to develop a higher bond-strength, quicker-
set adhesive from isocyanates. These chemical
companies were virtually looking for new markets for
their supply of isocyanates. The opposite trend may be
occurring among traditional suppliers of wood
adhesives. For phenolics, the wood adhesives market
represents a low profit, high volume segment. Many
companies are getting out of the adhesives market and
diverting their phenol to other value-added products.

With regard to R&D, trade associations and professional
societies serve as a conduit for information exchange
(see Exhibit 4.h). Through their regularly scheduled
programs and meetings, industrial research priorities
are passed on to government researchers and academi-
cians. The U.S. Department of Agriculture also
encourages this type of information dessemination. The
USDA Forest Products Laboratory sponsors an annual
meeting in the Fall for corporate representatives in
the wood products industry to discuss research needs
and priorities. A similiar meeting is held in the
spring with representatives from academia. Through
these forums, the Forest Products Laboratory sets up
its future research agendas. The National Forest
Products Association actively participates in these
forums each year.

Since information on overseas R&D activities was
limited to articles and only a few interviews, it is
more difficult to characterize the R&D network overseas. There are international societies promoting the
exchange of research findings in forestry. One such

organization is the International Union of Forest
Research Organizations. (See Exhibit 4.i for list of
countries and known areas of research.)

FACTORS INFLUENCING LEVEL OF R&D ACTIVITY

In the United States, the level of funding for R&D in
wood adhesives has always been lower than overseas.
Recently, the general level of R&D activity within the
industry at large has declined. Several factors
account for this decrease in R&D activity. The forest
products industry is in a depressed state, particularly
the wood panel producers. The industry was hard hit by
the 1982 recession. It suffered from the decline in
housing starts in the early eighties, and recently,
imported panel products have increased their share of
the domestic market. With a pattern similiar to the
domestic steel industry, the large wood producers are
selling off their older business segments and diversi-
fying into completely new types of business activities.
For example, Weyerhaeuser has moved into the real
estate and mortgage banking business and has sold its
resin manufacturing unit and laid off a large portion
of its R&D staff. The level of investment in forest
products R&D has suffered by this overall economic
decline in the industry.

Of the panel producers reported in Business Week's
annual summary of R&D expenditures by major industrial
groups, Boise Cascade, Weyerhaeuser and Masonite were
the only solid wood producers listed, the remaining
companies being primarily paper manufacturers. As
Exhibit 4.j shows, the level of R&D expenditures has
been erratic throughout the eighties, but the overall
trend has been downward. This decline has been corrobo-
rated by information supplied through interviews.
Corporate R&D managers in the forest products industry
noted tight budgets as the reason for less actvity in
wood adhesives R&D. For example, the R&D staff at the
Weyerhaeuser Technical Center in Tacoma, Washington
once comprised roughly 1,000 people. Today, the staff
has been dramatically cut in half.

During the energy crises of the seventies, resin
producers were literally cut off from their supply of
traditional petroleum feedstocks. Their ability to
service the wood adhesives market was at risk. Wood
panel producers became acutely aware of the volatility

Exhibit 4.i

Overseas R&D Activity by Country

R&D Players	Research in Alternative Wood Adhesives			
COUNTRY	**Lignin**	**Tannin**	**Isocyanate**	**Other**
Australia		X		
Brazil		X		
Canada	X			X
*Chile		X		
PRC/Taiwan		X		X
Federal Republic of Germany	X	X	X	X
Finland	X	X		
France	X	X		
India	X	X		X
Japan	X	X	X	X
New Zealand		X		
Nigeria		X		
Philippines	X			
South Africa	X	X		X
Soviet Union	X			X
Sweden	X	X	X	
*United Kingdom		X		

* Negligible level of R&D

Source: Hagler, Bailly & Company

Exhibit 4.j

R&D Expenditures for Select Wood Panel Manufacturers
(millions of dollars)

Wood Panel Producers	R&D Expenditures				
	1980	1981	1982	1983	1984
Boise Cascade	5.0	5.5	4.5	NA	7.9
Masonite	7.0	6.0	5.2	NA	NA
Weyerhaeuser	52.2	54.3	48.9	NA	45.6

NA = Information was not available

Source: Business Week

of their adhesives supply. Alternately, wood panel
producers have become increasingly reliant on adhesives
as the role of reconstituted wood products continues to
grow. Adhesives also dramatically affect the price of
wood products. The wood industry, except for specialized products, uses low cost adhesives to keep the
overall cost of their products down. Based on the
energy scare of the seventies, both resin manufacturers
and wood producers initiated programs in non-
conventional wood adhesives R&D.

With the recession in the eighties followed by an oil
glut, the level of corporate activity in non-
conventional adhesives has declined. It is clear that
the industry supports the efforts of academia in
pursuing non-conventional wood adhesives R&D. It is
viewed as their insurance policy against any future
supply disruptions. Also, industry recognizes that
over the long run, oil prices will increase as the
supply of oil becomes scarce. At that point it is
anticipated that continued R&D efforts in non-
conventional wood adhesives will reach a level of
practical feasibility.

5 Outlook

Of the three major research areas in non conventional bonding systems -- lignins, tannins, and isocyanates -- isocyanate is the most promising. Its potential for making a significant impact on the particleboard market within the next five years is great. In contrast, it is unlikely that any of the biomass-derived adhesives will penetrate the wood market in any substantial way within the next five years.

Isocyanate is a petroleum-derived polymer (see Exhibit 5.a for the properties of isocyanates and other adhesive systems). As such, it is a consistent and chemically clean product for formulation into an adhesive. Uniformity is an important industry specification for adhesives. Variability in products like lignins and tannins makes them less reliable. The failure of an adhesive system can lead to expensive legal proceedings over liability. Also, variability in a chemical derivative would result in constant adjustment of an adhesive mix to meet the technical specification of the mill. Constant adjustment would be too disruptive to the panel production process.

Other advantages to isocyanates are their high viscosity and shorter cure times. Isocyanate is a highly reactive chemical and can cure at room temperature. Shorter cure times can be achieved at higher temperatures. Because of its viscosity, less adhesive is used for bonding than with conventional adhesives. The tackiness of isocyanates is their major drawback. Without the aid of a release agent, isocyanates stick to the caul plates during pressing. Initially, this created a sufficient amount of down time in board production that further R&D was necessary to eliminate the problem altogether. Cheaper and better ways of averting this sticking problem are being developed and it appears that a practical solution is being attained.

With biomass-derived adhesive systems, if the scenario is one in which phenol, resorcinol, or urea are being directly substituted with modified lignins or tannins, variability is not the only drawback. Presently,

Exhibit 5.a
Summary of Plywood and Particleboard
Adhesive Technologies

Conventional Adhesive Systems	Technical Attributes				Economic Attributes		
	Reactivity	Bond Time	Temperature	Bond Strength	Source of Supply	Cost ($/lb)	Major End-use Application
I Phenol-formaldehyde (PF)	Moderate	Intermediate (high-temp)	High (180°F)	Strong	Oil	.32-.35	- Plywood - Exterior Particleboard
II Urea-formaldehyde (UF)	Moderate	Rapid (high-temp) Slow (room temp)	N/A	Fair	Natural	.16-.18	- Particleboard (Interior use)
III Melamine-formaldehyde (MF)	Moderate	Rapid (high temp)	N/A	Good	Natural Gas	.58-.60	- Plastics - Particleboard
IV Resorcinol-formaldehyde (RF)	High	Intermediate (room temp) Rapid (high temp)	Ambient	Strong	oil	$1.80	- Marine Plywood
Non-Conventional Adhesive Systems							
I Lignin	Low	Slow (9.5 min.)	High (180°F)	Fair	Wood	.40	- Plywood - Particleboard
II Tannin	High	Rapid	Ambient	Strong	Bark Nutshells	.35	- Marine Plywood
III Isocyanate	High	Rapid	Ambient	Strong	oil	N/A	- Particleboard

Source: Adapted from Skeist, I. Handbook of Adhesives, 1977.

lignins are a by-product of pulping. Once separated
from the kraft process waste stream, the lignins are
burned as an energy source in the chemical recovery
furnace. There has been some speculation as to how a
sufficient supply of lignin could be made available for
adhesive utilization. Some suggest that the improvements in the efficiency of the pulping process will
yield greater amounts of lignin. Others recommend
substituting coal for lignin as the energy source for
chemical recovery in kraft pulping. However, several
industry spokesmen have said that oil would be the
likely alternative based on easy handling.

With most lignin technologies, even if the technical
lignin requires further modification, the technology
can be linked directly to the pulping process.
Tannins, however, must first be extracted from bark.
This introduces a completely new stage of processing
unlinked to anything currently being done in the
industry. Adhesive utilization alone cannot justify
the costs of undertaking such an operation.

On the other hand, the prognosis for achieving tannin
utilization in adhesives appears brighter than with
lignin according to extensive industry interviews.
Chemically speaking, it is easier to slow down the
reactivity of tannins than to increase the reactivity
of lignin. As some have noted, only so many available
sites can be created on the lignin polymer to increase
its reactivity. To date, effective 100% replacement of
phenol with lignin has not been achieved and there is
evidence that significant deterioration of its adhesive
properties occurs beyond 75% substitution.

Tannins have been substituted for resorcinol in
adhesives. The main problem with tannins is producing
a cheap cold set resin consistently. A technical
lignin which is chemically modified is more consistent
than a tannin extract produced from bark.

While attempts to directly substitute lignin for phenol
have been impeded by slow cure and press times, there
is yet another approach to lignin utilization that has
proven to be technically feasible. This methodology
entails the the breaking down of lignin to cresols and
other phenolic precursors that could be used to
synthesize phenol. Lignin would become an alternate
feedstock for producing phenol. Where wood producers
have taken a dim view toward direct substitution of

natural substances for synthetic resins, those
questioned were receptive to the idea of breaking down
lignin for synthesizing phenol. A consistent chemical
product would be produced and variablity would be
overcome. Such a technology could be directly linked
to the pulping process. Phenol is relatively easy to
produce in a batch process which could be done at the
pulping facility. Only the economics of such an
approach is in question at the present time.

These trends are in sharp contrast to the developments
in non-conventional R&D overseas, but the drivers for
R&D here and abroad are very different. Energy prices
in Europe, Japan, South Africa, and Australia have been
higher than in the United States, where the availability of a domestic resource and price regulation
have kept energy prices artificially low. Investment
in non-conventional bonding systems has been greater
for a longer period of time overseas, where the supply
of phenol has been less dependable.

To a certain extent, the species of wood in other areas
of the world lend themselves better to adhesive
utilization. The North American species render fewer
phenolic groups than varieties available in South
Africa or Australia. Another factor encouraging R&D
overseas is the system of wood adhesive standards. In
the United States, testing standards have thwarted
efforts to introduce naturally derived adhesives. In
Europe, Japan, and elsewhere, the interpretation of
adhesion is related to strength values, but in the
United States, wood failure is a parameter for
evaluating bond quality. Most natural products achieve
good strength characteristics, but exhibit poor wood
failure.

As a result of these factors, more attempts have been
made overseas to introduce lignins or tannins as a
direct substitute for phenol and resorcinol. Many of
these technologies have been considered in the United
States, but have proven to be uneconomical. It is
unclear whether commercialization of non-conventional
bonding systems abroad has been accomplished for
reasons other than economics -- more stringent environmental regulations or the need for a more secure supply
-- or whether these technologies are truly economical
under differing market conditions.

R&D NEEDS FOR ALTERNATIVE ADHESIVES

There was a general consensus among wood adhesive R&D
players regarding research priorities and needs for the
development of lignins, tannins, and isocyanates as
alternatives to conventional thermosetting adhesives
(see Exhibit 5.b). A primary need is for fundamental
chemical investigation and characterization of native
lignins and tannins. Although this type of work has
been done overseas, the wood species available in the
United States are different and therefore require study
before they can be used. During the energy crisis,
attempts to import wattle tannins for use as alternative adhesive feedstock for wood adhesives proved
disastrous -- the adhesive performance of these wattle
tannin adhesives did not conform to U.S. manufacturing
practices.

Many of the problems associated with the failure of
earlier attempts to commercialize tannin or lignin
adhesive systems occurred within the molecular
structure of the natural substance. At the time,
little was known about the molecular composition of
various lignins and tannins and even less understood
was their interaction with other chemicals.

In the fifties and sixties several forest products
manufacturers tried to commercialize tannin extracts,
but after decades of work and millions of dollars
spent, these operations were shut down. Several
factors contributed to the downfall of these tannin
operations. At one level, there was a need to develop
other markets for the product besides wood adhesives.
Expanded markets and increased demand could possibly
justify the costs associated with operating a tannin
extraction plant. Also, the adhesive performance of
tannin-derived products was questionable. Failure in
the dimensional stability of the wood panels produced
with tannin adhesives was due in large part to the
inhibiting effect of the carbohydrate groups found in
the extract. At that time, information on the
molecular reactivity of tannins was virtually unknown.
As ongoing research reveals the structure and reactivity of these natural substances, procedures can be
developed to control their reactivity.

Another major R&D need has revolved around the issue of
formaldehyde emissions from conventional formaldehyde
adhesives and the implications for non-conventional

Exhibit 5.b

Major R&D Needs and Potential Energy Benefits (Losses)

Alternative Adhesive	Major R&D Need	Potential Energy Conservation Gains/Losses		
		Feedstock	Process	Total Bond Cost
Isocyanates	● Development of release agents	No change	Loss - complex multistage processing	Gain - faster setting resin
Lignins:				
- Direct substitution	● Increase reactivity ● Generate a uniform supply	Potential gain, although may need a replacement fuel source	Loss - purification and modification may be energy intensive	Loss - longer press times
- Breakdown to phenolic precursors	● Improve process efficiency to enhance economics	Potential gain, although may need a replacement fuel source	No change	No change
Tannins	● Slow down reactivity ● Generate a uniform product	Gain	Loss - extraction may be energy intensive	Gain - faster set resin

NOTE: Feedstock - refers to the embodied energies of the raw materials

Process - refers to the energy used in formulating the resin or isolating adhesive materials.

Total Bond Cost - refers to the energy costs associated with bonding, i.e., cure and press times, cure temperatures. These are the only energy costs directly born by the wood panel producer.

Source: Hagler, Bailly & Company, through data compiled from extensive industry interviews.

R&D. It has been shown that urea-formaldehyde under
humid conditions outgases formaldehyde. Urea-
formaldehyde is largely used for interior particleboard
applications where a cheap adhesive is required. Other
formaldehyde resins do not experience the same level of
formaldehyde release; however, it has been suggested
that substitues should be found for all formaldehyde
resins. Isocyanates provide this type of an alternative and in countries where formaldehyde use is
restricted, isocyanates have already been introduced
for certain particleboard applications.

While many researchers attempt to emulate the structure
and performance characteristics of formaldehyde-based
resins in developing non-conventional adhesives, it has
been suggested by some that an entirely new adhesive
system be constructed from non-conventional derivatives. One approach may be the formulation of aldehyde
from furfuryl which can then be reacted to a modified
lignin. Aldehyde would behave in much the same way as
formaldehyde does in a conventional adhesive system;
however, aldehyde would not have the toxic effects of
formaldehyde.

CONCLUSIONS

The wood panel industry in the United States and other
industrialized countries is undergoing significant
change. In many countries the timber base is declining
or as in the U.S., is changing substantially. The
replacement of high quality trees is not rapid enough
to meet the continuing demand for wood products.
Consequently, younger, poorer quality trees must be
used for panel products. Because of this declining
wood base, and even more importantly because of
economic pressures to maintain competitive prices for
wood products, reconstituted wood products are growing
and expanding into areas once reserved for solid wood
products. The performance of these reconstituted wood
products is highly dependent on adhesives. In today's
panel products, the strength of the panel can be
greater than the individual wood components because of
the strength of the adhesive system. As a result,
adhesives are becoming a larger and more important
component of the end-product.

There are general R&D needs that arise from the
changing requirements of the wood industry based on the
state of the current timber base. Panel products are

being manufactured from wood with a higher moisture
content. Therefore, future adhesives must be able to
adhere to wood with a higher moisture content. If not,
total bond costs will rise because of the increased
energy costs associated with extended drying of the
wood in preparation for bonding. Also, trees of poorer
quality are more difficult to finish. Therefore, new
bonding systems must have the capability to adhere to
rough surfaces.

As the wood industry becomes increasingly dependent on
chemical components and advanced means of processing
wood residues, adhesive material costs and energy costs
will become more critical to their operations. The
industry has already become concerned with cure and
press times. Adhesives with shorter cure and press
times along with lower cure temperatures will be
increasingly important to the industry.

Clearly, there is a need for continued R&D in non-
conventional bonding systems. Eventually,
petrochemical supplies may not be able to meet the
growing need for adhesives by the wood panel industry
or the prices of petroleum-based adhesives may become
too high. At this point, natural substances will be an
alternative to petroleum-based adhesives. Continued
research into the fundamental chemistry of these
natural substances will make it easier to put in place
an economical and technically feasible alternative
adhesive system. The level of R&D currently taking
place in wood adhesives is limited. Without this R&D,
when it does become necessary to make a transition to
non-petroleum based adhesives, an adequate information
base may not be at hand to make the proper decisions.

In the near term, isocyanates are likely to satisfy any
immediate needs for an adhesive system that is not
founded on formaldehyde. However, isocyanates are a
petroleum-based product and would not leave the wood   *use less*
industry in a better position to secure its adhesive
supply from any future oil shocks. Therefore, it is
critical for research on biomass-derived adhesives to
continue.

Many trade-offs exist between lignin and tannin
utilization for wood adhesives. It is likely that both
will have a future role in the wood adhesives market,
although to what extent either will penetrate the
market or be used in specific end-use applications

remains unclear. It is certain that further research
into the fundamental chemical characteristics of
lignins and tannins is necessary before either can
successfully reach commercialization.

Through fundamental research, the reactivity of both
lignins and tannins may be effectively controlled. By
controlling reactivity, it is likely that these natural
substances can be modified to meet industry specifications. The key specifications industry maintains for
adhesive utilization are the following:

●   Consistency -- as relates to the reliability of
    the bond. Usually requires a chemically clean
    product so that an adhesive mix will have
    predictable performance characteristics.

●   Cost -- a factor since the industry must keep down
    the price of its panel products by using inexpen-
    sive adhesives. There are only a limited number
    of end-use applications where the consumer will
    pay a premium for better performance.

●   Level of panel production -- the industry will not
    tolerate longer cure or press times because
    extended cure or press times can significantly
    impact panel production.

●   Equipment changes -- equipment changes are costly
    to the wood panel industry, and since it is a
    depressed industry, few if any equipment changes
    will be tolerated. This has implications for both
    adhesive production equipment and adhesive
    application and press equipment.

The potential energy impacts of successful alternative
wood adhesives could be significant. (See Exhibit 5.b
for assessment of energy gains/losses through
alternative adhesive use.)

It is likely that some or all of the alternative
adhesives will enter the market in the next 5-10 years.
Much fundamental research must first be carried out to
characterize domestic biomass resources and to develop
an understanding of how to control their reactivity and
form a dependable adhesive bond. However, once these
technical and economic problems have been solved,
biomass-derived adhesives for the wood panel industry
could play an important role in conserving scarce U.S.

fossil fuel resources by substituting renewable biomass for current fossil fuel-based raw materials.

Appendix A: Tables of Market Data

Exhibit A.1

U.S. Market Parameters for Resin Feedstocks (1982)

$(10^6 lb)$

	Production	Demand	Capacity	Imports	Exports	Markets	End Uses % of Production
Ammonia	30,980	33,720	39,520	4,220	1,480	Fertilizer materials fiber, plastic, resin intermediates explosives miscellaneous	80 11 4
Benzene	7,841	8,780	17,600	998	59	cumene ethylbenzene (styrene) cyclohexane (nylon) nitrobenzene (aniline) alkylbenzene (detergents) chlorobenzene miscellaneous	20 50 15 5 3 2 5
Cumene	2,743	2,893	4,600	171	21	phenol and acetone miscellaneous	99 1
Formaldehyde	4,691	4,681	8,695	--	10	urea formaldehyde resins phenol formaldehyde resins melamine formaldehyde resins pentaerythritol 1,4-butane diol acetal resins hexamethylenetetramine methyl diphenyl diisocyanates miscellaneous	31 22 4 9 9 8 6 2 9
Methanol	7,525	6,766	10,470	357	1,116	formaldehyde acetic-acid chloromethanes methyl methacrylate methylamines dimethyl terephthalate automotive chemicals and solvents other non-fuel uses gasoline blending methyl tertiaryu butyl ether	33 14 6 3 3 3 9 16 8 5
Melamine	80	102	145	22	--	laminating resins	26

Exhibit A.1 (continued)

U.S. Market Parameters for Resin Feedstocks (1982)

	Production	Demand	Capacity	Imports	Exports	Markets	End Uses % of Production
			$(10^6 lb)$				
						surface coating resins	24
						molding compounds	18
						paper treatment resins	12
						textile treatment resins	8
						adhesives and bonding	6
						miscellaneous	6
Phenol	2,175	2,064	3,085	--	111	phenolic resins	45
						bisphenol A	25
						caprolactam	15
						alkylphenols	5
						miscellaneous	10
Resorcinol	28	20	40-45	2	10	rubber products	68
						wood adhesives	11
						miscellaneous	21
Urea	15,224	13,772	15,560	1,706	3,158	fertilizers	95
						animal feeds	2
						miscellaneous and urea- formaldehyde resins	3

Exhibit A.2
U.S. Thermosetting Adhesive Producers

Company	Resin[1]					
	PF	PF/RF	RF	UF	UF/MF	MF
Adhesive Products Corporation	X	X	X	X		X
American Cyanamid Company				X	X	X
American Finish & Chemical Company						X
American Glue & Resin Company				X		
ArChem Corporation		X		X		X
Baltimore Adhesive Company				X		
Borden Chemical	X	X	X	X		X
Bostik Division, Emhart Corporation	X	X	X			
W.H. Brady Technicote Division				X		
Cadillac Plastic & Chemical Company, Division of Dayco Corporation	X					
The Cambridge Tile Mfg. Company		X		X	X	
Can-Tite Rubber Corporation	X	X	X	X	X	X
R.H. Carlson Company, Inc.			X			
Cascade Resins, Inc.			X			
Chembond Corporation	X			X	X	X
Clifton Adhesive, Inc.			X			
Dolphin Paint & Chemical Company				X		X
Dunbar Sales Company, Inc.	X	X	X	X	X	
Thomas W. Dunn Corporation				X		
Dural Products, Ltd.				X		
Electro-Kinetic Systems	X					
Finetex, Inc.				X		
H.B. Fuller Company - Puerto Rico	X					
Fuller Paste & Adhesives Company			X			
GPM Inc.	X	X		X	X	
General Electric Company, Silicone Products Division			X			
Georgia-Pacific Corporation	X	X		X	X	X
Gossett Resins & Chemicals, Inc.	X	X	X	X	X	X
Gulf Adhesives & Resins	X	X	X	X	X	X
C.L. Hauthaway & Sons Corporation	X	X		X	X	X
Helmitin Canada Inc.	X			X		
Hermetite Products Inc.			X			
Industrial Adhesives, Inc.	X					
Ironsides Resins Division, Ironsides Company	X	X				
Key Polymer Corporation	X	X				X
Koppers Company		X	X			
LePage's Limited				X		
Loctite Corporation			X			
Mearl Corporation, Foam & Chemical Division				X		
Monsanto Polymer Products Company				X	X	X

Exhibit A.2 (continued)
U.S. Thermosetting Adhesive Producers

Company	Resin[1]					
	PF	PF/RF	RF	UF	UF/MF	MF
National Starch & Chemical Corporation	X					
Nicholson & Company, Division of Nico, Inc.					X	
Northeast Chemical Company, Inc.				X	X	
Occidental Chemical Corporation	x					
Pacific Resins & Chemicals, Inc.	X	X		X		
Passaic Adhesive & Chemical Company				X		
Pilot Chemical Company, Inc.			X	X		
Polymerics, Inc.					X	X
Polyrez Company, Inc.	X					
Raymark Corporation	X					
Refined Onyx Company, Millmaster Onyx Group	X			X	X	X
Reichhold Chemicals, Inc.	x			x		x
Research Sales, Inc.	X			X	X	
J. E. Rhoads & Son, Inc.			X			
Roberts Consolidated Industries, Inc.			X	X		
Roman Adhesives, Inc.			X			X
Schenectady Chemicals, Inc.	x					
Sentry/Custom Services Corporation	X			X	X	X
Southeastern Adhesives, Inc.				X		
Spartan Adhesives & Coatings Company	X	X	X			
Specialty Coatings & Chemicals, Inc.		X				
Technical Coatings Laboratory	X			X	X	X
Union Carbide	X					
Valchem Chemical Division, United Merchants				X	X	X
Valley Adhesives & Coatings	X					X
Waterlac Industries, Inc.				X		
West Coast Adhesives	X	X	X	X	X	X
Western American, Inc. dba Western Adhesives				X		
Weyerhaeuser Company	X	X				
Wilhold Glues, Inc.		X				

[1]PF = Phenol-formaldehyde
PF/RF = Phenol-formaldehyde/resorcinol formaldehyde
RF = Resorcinol-formaldehyde
UF = Urea-formaldehyde
UF/MF = Urea-formaldehyde/Melamine-formaldehyde
MF = Melamine-formaldehyde

Sources: Adhesives Age, 1985
          Hagler, Bailly & Company

Exhibit A.3

U.S. Production of Phenolic, Resorcinol, Urea, and Melamine Resins
(10^6 lb dry weight)

| Year | Resin | | | |
	Phenolic[1]	Resorcinol	Urea	Melamine
1973	1,387	NA	867	170
1974	1,335	NA	835	164
1975	1,051	27	690	115
1976	1,340	31	821	186
1977	1,458	35	963	200
1978	1,617	36	1,122	202
1979	1,781	38	1,367	200
1980	1,499	32	1,165	167
1981	1,280	31	1,165	179
1982	1,100	28	998	158
1983	1,320	31	1,172	180
1984	1,430	NA	1,199	210

[1]Includes other tar acid resins.

NA = Not available.

Source: Chemical & Engineering News; U.S. Department of Commerce;
Chemical Marketing Reporter.

Exhibit A.4

U.S. Resorcinal Imports

	Quantity (lbs)
Canada	2,646
Belgium	827
France	1,000
Germany (Federal Republic of)	571,104
Japan	1,371,793
TOTAL	**1,947,370**

Source: U.S. Department of Commerce

Exhibit A.5

Average Variable Costs for U.S. Southern Plywood Mills

($/MSF, 3/8-INCH BASIS)

	Annual											
	1976	1977	1978	1979	1980	1981	1982	1983	1984	1985	1986	1987
WOOD												
STUMPAGE	41	51	59	72	73	73	69	67	65	63	64	62
HARVEST	16	17	19	20	21	22	22	24	25	25	26	27
TOTAL WOOD COST	57	68	78	92	94	95	91	91	90	88	90	89
LABOR(DIRECT,INDIRECT. & MAINTENANCE)	22	25	29	31	31	34	34	34	35	35	37	37
ENERGY	8	9	9	10	11	13	14	14	15	16	16	17
GLUE	9	8	7	9	11	11	11	10	11	11	12	12
MISCELLANEOUS	11	12	13	13	14	14	13	12	12	11	13	14
GROSS VARIABLE COSTS	107	122	136	155	161	167	163	161	163	163	167	169
RESIDUES	14	15	18	21	24	25	25	25	27	26	25	28
NET VARIABLE COSTS (GROSS LESS RESIDUES)	93	107	118	134	137	142	138	136	136	137	142	148
GENERAL & ADMINISTRATIVE	6	6	7	7	8	7	7	8	8	8	8	8
TOT. HIGH VAR. COSTS(3/8-INCH)	99	113	125	141	145	149	145	144	144	146	150	153
TOT. HIGH VAR. COSTS(1/2-INCH)	132	151	167	188	193	199	193	192	193	194	200	204

	Annual											
	1976	1977	1978	1979	1980	1981	1982	1983	1984	1985	1986	1987
STUMPAGE-BID PRICE	..	118	146	183	164	183	165	178	172	181	171	198
STUMPAGE-HARVEST PRICE	97	119	137	173	182	174	171	177	176	173	178	176
RECOVERY RATE	2.370	2.330	2.330	2.400	2.420	2.400	2.480	2.635	2.700	2.750	2.800	2.820
PLYWOOD TALLY STUMPAGE	41	51	59	72	73	73	69	67	65	63	64	62

NOTES:
(1) STUMPAGE AND HARVEST COSTS DERIVED FROM LA. QTRLY. REPTS. PRIOR TO 1976, AND SUBSEQUENTLY FROM TIMBER MART SOUTH.
TIMBER MART SOUTH DATA FOR 10 STATES(ALA.,ARK.,FLA.,GA.,LA.,MISS,N.C.,S.C.,TX.,VA.)ARE WEIGHTED BY VOLUME TO DERIVE
SOUTHERN STUMPAGE AND DELIVERED COSTS. REMAINDER ARE FORSIM FORECASTS.
WHERE:(A)PLYWOOD TALLY STUMPAGE=HARVEST PRICE/RECOVERY RATE
(B)HARVEST PRICE=WEIGHTED AVG. FOR PRECEDING BID PRICES.
(C)ANNUAL STUMPAGE BID PRICES ARE UNWEIGHTED AVGS.OF QTRLY. DATA(D)ALL RECOVERY FACTORS ON SCRIBNER SCALE BASIS
(3) HISTORICAL LABOR COSTS ARE CALCULATED BY THE APA,AS ARE ENERGY,GLUE,MISCELLANEOUS COSTS AND RESIDUE INCOME DATA.
(4) GENERAL AND ADMINISTRATIVE EXPENSES ARE BASED UPON APA DATA. FORECASTS ARE ESTIMATED AS 4.5% OF GROSS VARIABLE COSTS.
(5) SOUTHERN YELLOW PINE HIGH AVERAGE VARIABLE COSTS APPLY TO MILLS PURCHASING PEELER LOGS ON THE OPEN MARKET. WOOD COSTS
FOR MILLS OPERATING OFF TIMBER OWNED BY THE MILL WOULD BE MUCH LOWER.

Source: Data Resources, Inc.

Exhibit A.6

Average Variable Costs for U.S. North Central Waferboard Mills

($/MSF, 3/8-INCH BASIS)

	Annual											
	1976	1977	1978	1979	1980	1981	1982	1983	1984	1985	1986	1987
WOOD (DELIVERED)	19.75	19.38	20.63	22.49	23.98	24.58	25.55	26.51	27.94	29.18	31.25	32.50
LABOR	7.75	8.25	8.80	9.41	10.33	11.49	12.61	12.69	12.55	12.66	13.03	13.31
ENERGY	3.09	3.53	3.87	4.61	6.00	7.39	7.92	8.12	8.31	8.34	8.60	9.00
GLUE	13.26	14.18	15.22	17.18	19.16	20.33	19.98	16.97	16.85	17.19	16.95	17.59
WAX	2.76	2.96	3.17	3.58	4.17	4.61	4.73	4.79	4.96	5.05	5.24	5.44
MISC.	8.02	8.58	9.21	10.40	12.08	13.37	13.73	13.91	14.39	14.64	15.19	15.76
GROSS VARIABLE COSTS	54.64	56.89	60.90	67.67	75.72	81.78	84.52	82.99	84.99	87.06	90.26	93.60
GENERAL & ADMINISTRATIVE	2.73	2.84	3.04	3.38	3.79	4.09	4.23	4.27	4.25	4.35	4.51	4.68
TOTAL VARIABLE COSTS	57.37	59.73	63.94	71.06	79.51	85.87	88.75	87.26	89.24	91.41	94.77	98.28
% CHANGE	6.7	4.1	7.0	11.1	11.9	8.0	3.4	-1.7	2.3	2.4	3.7	3.7

NOTES:
(1) MAIN SOURCE: DATA RESOURCES, INC. AND INDUSTRY SURVEYS.
(2) PRICE AND CONVERSION ASSUMPTIONS USED TO DERIVE THE WOOD, LABOR, ENERGY AND GLUE COSTS ARE OUTLINED IN THE CORRESPONDING TABLE ENTITLED "FACTOR INPUT PRICE AND CONVERSION ASSUMPTIONS".
(3) WAX, MISC. AND GENERAL & ADMINISTRATIVE COSTS ARE FORECAST USING COST ESCALATOR INDICES FROM THE DRI U.S. ECONOMIC SERVICE.
(4) GROSS VARIABLE COSTS EQUAL THE SUM OF WOOD, LABOR, ENERGY, GLUE, WAX AND MISC. EXPENSES.
(5) TOTAL VARIABLE COSTS EQUAL THE SUM OF GROSS VARIABLE AND GENERAL & ADMINISTRATIVE COSTS.

Source: Data Resources, Inc.

Exhibit A.7

Average Variable Costs for U.S. North Central Oriented Standardboard Mills

($/MSF, 3/8-INCH BASIS)

	Annual											
	1976	1977	1978	1979	1980	1981	1982	1983	1984	1985	1986	1987
WOOD (DELIVERED)	20.79	20.40	21.71	23.67	25.24	25.87	26.89	27.90	29.41	30.71	32.89	34.21
LABOR	7.75	8.25	8.80	9.41	10.33	11.49	12.61	12.84	12.55	12.66	13.03	13.31
ENERGY	3.16	3.61	3.96	4.71	6.11	7.52	8.06	8.28	8.47	8.51	8.77	9.18
GLUE	17.69	18.92	20.30	22.54	25.30	27.03	24.63	23.94	24.36	23.88	23.72	24.07
WAX	1.63	1.74	1.87	2.11	2.45	2.72	2.79	2.83	2.93	2.98	3.09	3.21
MISC.	8.02	8.58	9.21	10.40	12.08	13.37	13.73	13.91	14.39	14.64	15.19	15.76
GROSS VARIABLE COSTS	59.03	61.50	65.84	72.83	81.52	88.00	88.73	89.69	92.10	93.37	96.70	99.75
GENERAL & ADMINISTRATIVE	2.95	3.08	3.29	3.64	4.08	4.40	4.44	4.45	4.61	4.67	4.84	4.99
TOTAL VARIABLE COSTS	61.98	64.58	69.14	76.47	85.60	92.40	93.16	94.14	96.71	98.04	101.54	104.73
% CHANGE	--	4.2	7.1	10.6	11.9	8.0	0.8	1.0	2.7	1.4	3.6	3.1

NOTES:
(1) MAIN SOURCE: DATA RESOURCES, INC. AND INDUSTRY SURVEYS.
(2) PRICE AND CONVERSION ASSUMPTIONS USED TO DERIVE THE WOOD, LABOR, ENERGY AND GLUE COSTS ARE OUTLINED IN THE CORRESPONDING TABLE ENTITLED "FACTOR INPUT PRICE AND CONVERSION ASSUMPTIONS".
(3) WAX, MISC. AND GENERAL & ADMINISTRATIVE COSTS ARE FORECAST USING COST ESCALATOR INDICES FROM THE DRI U.S. ECONOMIC SERVICE.
(4) GROSS VARIABLE COSTS EQUAL THE SUM OF WOOD, LABOR, ENERGY, GLUE, WAX AND MISC. EXPENSES.
(5) TOTAL VARIABLE COSTS EQUAL THE SUM OF GROSS VARIABLE AND GENERAL & ADMINISTRATIVE COSTS.

Source: Data Resources, Inc.

Appendix B: References

1. Abe, I. "Some of the Japanese Researches in Natural Adhesives for Wood", International Union of Forest Research Organizations, World Congress Proceedings, 1981, pp. 177-183.

2. Agoos, Alice. "Phenol: Tight Supplies, Weak Prices", Chemical Week, May 8, 1985, pp.24-26.

3. "Amino Resins and Plastics", Kirk-Othmer Concise Enclyclopedia of Chemical Technology, Wiley-Interscience, 1985, pp.90-92.

4. Austin, George T. Shreve's Chemical Process Industries, 5th Edition, McGraw-Hill, 1984.

5. Ball, G.W. "New Opportunities in Manufacturing Conventional Particleboard Using Isocyanate Binders", International Symposium on Particleboard, Washington State University, 1981, pp. 265-285.

6. "Bark Adhesive Plant", Australian Forest Industries Journal, October 1981, p. 25.

7. Bauer, Eric. "Recent Advances in Plywood Technology", Adhesives for Wood, edited by Robert Gillespie, Noyes Publications, 1984.

8. Blackman, Ted. "New Process Goes Commercial", World Wood, March 1980, pp. 12-13.

9. Brink, D.L., M.L. Kuo, W.E. Johns, M.J. Birnbach, H.D. Layton, T. Nguyen, and T. Breiner. "Exterior Particleboard Bonded with Oxidative Pretreatment and Crosslinking Agent", Holzforschung, Vol. 37, No. 2, 1983, pp. 69-78.

10. Brown, Harry L., Bernard B. Hamel, and Bruce A. Hedman. Energy Analysis of 108 Industrial Processes, Fairmont Press, 1985.

11. Business Week, McGraw-Hill Publication, 3/21/84; 7/8/85.

12. Cameron, F.A., and A. Pizzi. "Acid-Setting Cold-Setting Phenolic Adhesives for Glulam: A Controversial Issue", Journal of Applied Polymer Science: Applied Polymer Symposium, 1984, pp. 229-234.

13. Cameron, F.A., and A. Pizzi. "Phenolic and Tannin-Based Honeymoon-Type Adhesives: (I) Glulam and Finger Joints Application and (II) Minimum Curing Temperatures", Wood Industry Abstracts, Vol. 13, No. 6, November 1984, p. 689.

14. Carll, Charles G., H. Edward Dickerhoof, and John A Youngquist, "U.S. Wood-Based Panel Industry", U.S. Department of Agriculture, Forest Service, 1982.

15. Chemcyclopedia 86, American Chemical Society, 1985.

16. Chemical Products Synopsis, Mannsville Chemical Products, July 1, 1983-June 30, 1984.

17. Chen, Chia M. "Effects of Extraction on the Quantity of Formaldehyde Requirement in Reaction of Extractives of Bark and Agricultural Residues with Formaldehyde", Holzforschung, Vol. 36, No. 2, 1982, pp. 65-70.

18. Chen, Chia M. "The Use of Forest and Agricultural Residue Extracts in the Production of Exterior Phenolic Resin Adhesives", Adhesives for Wood, edited by Robert Gillespie, Noyes Publications, 1984.

19. Chow, Poo, Gary L. Rolfe, and Lester E. Arnold. "Chemicals, Fiber, and Energy from Woody Biomass", Illinois Research, Winter 1983, pp. 11-13.

20. Chow, Poo, and Marvin P. Steinberg. "The Quest for an Efficient Conversion Process", Illinois Research, Spring/Summer 1984, pp. 8-11.

21. Chow, S. "Adhesive Developments in Forest Products," Wood Science and Technology, No. 17, 1983, pp. 1-11.

22. Christiansen, Alfred W. "Acid-Catalyzed Phenolic Bonding of Wood", Adhesives for Wood, edited by Robert Gillespie, Noyes Publications, 1984.

23. Chum, H.L., S.K. Parker, D.A. Feinberg, J.D. Wright,
     P.A. Rice, S.A. Sinclair, and W.G. Glasser. The
     Economic Contribution of Lignins to Ethanol
     Production from Biomass, Solar Energy Research
     Institute, May 1985.

24. Coppens, H.A., M.A.E. Santana, and F.J. Pastore.
     "Tannin Formaldehyde Adhesive for Exterior-Grade
     Plywood and Particleboard Manufacture", Forest
     Products Journal, April 1980, pp. 38-42.

25. Data Resources, Inc. FORSIM, 1985.

26. Dickerhoof, H.E. et al.   "U.S. Wood-Based Panel
     Industry: Parts I-IV." Forest Products Journal,
     reproduced by USDA Forest Service, 1982.

27. Ebewele, R.O., O.A. Peters, and J.Y. Olayemi.
     "Development of Wood-Products Adhesives from Mangrove
     Bark", Journal of Applied Polymer Science, vol. 29,
     1984, pp. 1415-1426.

28. Energy and Environmental Analysis, Inc. Industrial
     Energy Productivity Project - Final Report: The
     Chemicals Industry, DOE/CS/40151-1, U.S. Department
     of Energy, February 1983.

29. Environmental Protection Agency.   Development
     Document for Effluent Limitations Guidelines and
     Standards for the Pulp, Paperboard and the Builder's
     Paper and Board Mills, December 1980.

30. FAO, Timber Bulletin for Europe, V.N. Publications,
     January-June 1982.

31. FAO, World Forest Products Demand and Supply 1990 and
     2000, 1982.

32. FAO, Yearbook of Forest Products 1982, U.N.
     Publications.

33. Gamo, Masato. "Wood Adhesives from Natural Raw
     Materials", Journal of Applied Polymer Science:
     Applied Polymer Symposium, 1984, pp. 101-126.

34. Gaul, J.M., T. Nguyen, and J.S. Babiec, Jr. "Novel
     Isocyanate Binder Systems for Composite Wood Panels",

Journal of Elastomers and Plastics, July 1984, pp. 206-228.

35. Gellerstedt, Goran, and Eva-Lisa Lindfors. "Structural Changes in Lignin During Kraft Pulping", *Holzforschung*, Vol. 38, no. 3, 1984, pp. 151-158.

36. Gendler, J.L., D.T.A. Huibers, and H.J. Parkhurst, Jr. "Hydroxy-Aromatics from Lignin Hydrogenolysis", *Wood and Agricultural Residues*, Texas A&M University, Department of Forest Science, 1983, pp. 391-399.

37. George, J., S.S. Zoolagud, D.M. Raghunath Rao, T.R. Narayanaprasad, K.K. Mohandas, and T.S. Rangaraju. "Tannin Adhesives for Plywood", *IPIRI Journal*, Vol. 6, No. 2, pp. 69-73.

38. Gillespie, Robert H. ed. *Adhesives for Wood: Research, Applications and Needs*, Noyes Data Corp., 1984.

39. Gillespie, Robert H. "Bonded Wood Products -- A Review", *Journal of Adhesion*, Vol. 15, 1982, pp. 51-58.

40. Gillespie, Robert H. "Lignin in Durable Adhesives for Wood: Part I. Evaluation of Lignin-Aldehyde Reactivity", *FPRS Adhesives Symposium Proceedings*, U.S. Department of Agriculture, Forest Service, October 1985.

41. Glasser, Wolfgang G., Leo C.-F. Wu, and Johan-F. Selin. "Synthesis, Structure, and Some Properties of Hydroxypropyl Lignins", *Wood and Agricultural Residues*, Texas A&M University, Department of Forest Science, 1983, pp. 149-165.

42. Gollob, L., R.L. Krahmer, J.D. Wellons, and A.W. Christiansen. "Relationship Between Chemical Characteristics of Phenol-Formaldehyde Resins and Adhesive Performance", *Forest Products Journal*, March 1985, pp. 42-48.

43. Graff, Gordon M. "High-Grade Lignin Schemes, Edge Closer to Reality," *Chemical Engineering*, 12/27/82, pp. 25-27.

44. Griggs, B.F., L-W. Zhao, C-L. Chen, and J.S. Gratzl. "Sulfomethylation and Oxidative Sulfonation of Kraft Lignin", Department of Wood and Paper Science, North Carolina State University.

45. Grozdits, G.A., and J.N. Bibal. "Surface-Activated Wood Bonding Systems: Accelerated Aging of Coated and Uncoated Composite Boards", Holzforschung, Vol. 37, No. 4, 1983, pp. 167-172.

46. Gutcho, Marcia, ed. Adhesives Technology: Developments Since 1979, Noyes Data Corp., 1983.

47. Hall, J. Alfred. Utilization of Douglas Fir Bark, USDA, GPO 1971.

48. Haynes, Richard W. and Marius M. Adams. "Matching Projections of Supply and Demand for Forest Products in the U.S.", Forest Products Journal, October 1981, pp. 77-81.

49. Herrick, Franklin W. "Chemistry and Utilization of Western Hemlock Bark Extractives", Journal of Agriculture and Food Chemistry, No. 28, 1980, pp. 228-237.

50. Holowacz, J. "Forests of the USSR", The Forestry Chronicle, Vol. 61, No. 5 1985, pp. 366-373.

51. Houwink, R.; Salomon, G. Adhesion and Adhesives, Volume 2, Elsevier Publishing Co., 1967.

52. Huber, A.E., A. Pizzi, and F.A. Cameron. "Industrial Tannin Adhesives for Interior Plywood", Journal of Applied Polymer Science: Applied Polymer Symposium, 1984, pp. 59-61.

53. "Hydroquine, Resorcinol and Catechol", Kirk-Othmer Encyclopedia of Chemical Technology, Vol. 13, 1981, pp.39-69.

54. Johns, William E. "Is There an Isocyanate in Your Future?", International Symposium on Particleboard, Washington State University, 1980, pp. 177-184.

55. Industrial Energy Use Book, Oak Ridge Associated Universities, ORAU-160, 1980.

56. Johns, William E., H.D. Layton, T. Ngyuen, and J.K. Woo. "The Nonconventional Bonding of White Fir Flakeboard Using Nitric Acid", <u>Holzforschung</u>, Vol. 32, No. 5, 1978, pp. 162-166.

57. Johns, William E., T.M. Maloney, E.M. Huffaker, J.B. Saunders, and M.T. Lentz. "Isocyanate Binders for Particleboard Manufacture", <u>International Symposium on Particleboard</u>, Washington State University, 1981, pp. 213-221.

58. Johns, William E, and Tinh Nguyen. "Peroxyacetic Acid Bonding of Wood", <u>Forest Products Journal</u>, January 1977, pp. 17-23.

59. Jse, Chung Yun. "Effect of Resin Types and Formulation in Internal Bond Strength and Dimensional Stability of Hardwood Flakeboard", <u>Adhesives for Wood</u>, edited by Robert Gillespie, Noyes Publications, 1984, pp. 40-49.

60. Kelley, S.S., R.A. Young, R.M. Rammon, and R.H. Gillespie. "Bond Formation by Wood Surface Reactions: Part III -- Parameters Affecting the Bond Strength of Solid Wood Panels", <u>Forest Products Journal</u>, February 1983, pp. 21-28.

61. Kelley, S.S., R.A. Young, R.M. Rammon, and R.H. Gillespie. "Bond Formation by Wood Surface Reactions: Part IV -- Analysis of Furfuryl Alcohol, Tannin and Maleic Acid Bridging Agents", <u>Journal of Wood Chemistry and Technology</u>, Vol. 2, No. 3, 1982, pp. 317-342.

62. Kent, James A., ed. <u>Riegel's Handbook of Industrial Chemistry</u>, 8th ed., Van Nostrand Reinhold Company, 1983.

63. Kirkman, A.G., L.L. Edwards, and J.S. Gratzl. "Kraft Lignin Recovery by Ultrafiltration -- Economic Feasibility and Impact on Kraft Recovery System", <u>Journal Series of the North Carolina Agricultural Research Service</u>, Paper No. 9348.

64. <u>Kirk-Othmer Concise Encyclopedia of Chemical Technology</u>, John Wiley & Sons, 1985.

65. Koch, Peter. "Material Balances and Energy Required for Manufacture of Ten Wood Commodoties", Energy and the Wood Products Industry, Forest Products Research Society, 1976.

66. Kondo, Ryuichiro, and Joseph L. McCarthy. "Incremental Delignification of Hemlock Wood and Characterization of Lignin Products", Holzforschung, Vol. 39, No. 4, pp. 231-234.

67. Kreibich, Roland E., and Richard W. Hemingway. "Condensed Tannin-Resorcinol Adducts in Laminating Adhesives", Forest Products Journal, March 1985, pp. 23-25.

68. Kruzas, A.T. et al. International Research Centers Directory, Gale Research Co., 1982.

69. Kruzas, A.T. and K. Gill, eds. International Research Centers Directory, Gale Research Company, 1982.

70. Krzysik, Andrzej, and Raymond A. Young. "A Lignin Adhesive System for Flakeboard Production", Submitted to Forest Products Journal, 1985.

71. Kumar, Adarsh, R.C. Gupta, and N.C. Jain. "A Preliminary Note on Plywood Adhesives from Bark Tannins of Sal (Shorea Robusta)", IPIRI Journal, Vol. 4, No. 3, pp. 126-128.

72. Lee, Lieng-Huang. Adhesive Chemistry: Developments and Trends, Plenum Press, 1984.

73. Leitheiser, Robert H., Ben R. Bogner, and Frank C. Grant-Acquah. "Water Dilutable Furan Resin Binder for Particleboard", Adhesives for Wood, edited by Robert Gillespie, pp. 70-76.

74. "Lignin Is Coming of Age for Use in Polymeric Materials", Chemical & Engineering News, September 24, 1984, pp. 19-20.

75. Lipinsky, Edward S. "Disruption and Fractionation of Lignocellulose", Wood and Agricultural Residues, Texas A&M University, Department of Forest Science, 1983, pp. 489-501.

76. Marchessault, R.H., S.L. Malhotra, A.Y. Jones, and A. Perovic, "The Wood Explosion Process: Characterization and Uses of Lignin/Cellulose Products", Wood and Agricultural Residues, Texas A&M University, Department of Forest Science, 1983, pp. 401-413.

77. Marra, Alan A. "Adhesives for Wood Composites", Adhesives for Wood, edited by Robert Gillespie, Noyes Publications, 1984, pp. 223-227.

78. Marra, George G. "The Role of Adhesion and Adhesives in the Wood Products Industry", Adhesives for Wood, edited by Robert Gillespie, Noyes Publications, 1984, pp. 2-9.

79. McKean, William T., ed. Energy and Environmental Concerns in the Forest Products Industry, AICLE Symposium Series, Vol. 74, No. 177, 1978.

80. McKeever, David B., and Cherilyn A. Hatfield. Trends in the Production and Consumption of Major Forest Products in the United States, Resource Bulletin FPL 14, U.S. Department of Agriculture, Forest Products Laboratory, July 1984.

81. McLaughlin, Alexander, Louis M Alberino, William J. Farrissey, and Douglas P. Waszeciak. "Polymeric Isocyanate as Wood Product Binders", Adhesives for Wood, edited by Robert Gillespie, Noyes Publications, 1984, pp. 130-149.

82. Mohandas, K.K., T.R. Narayanaprasad, and Joseph George. " Maize Gluten as a Water Resistant Extender for Synthetic Resin Adhesives, Part III: Hot Press Phenol-Formaldehyde Resin Adhesives", IPIRI Journal, Vol. 6, No. 2, pp. 62-65.

83. Muller, Peter C., and Wolfgang G. Glasser. "Engineering Plastics from Lignin. VIII. Phenolic Resin Prepolymer Synthesis and Analysis", Journal of Adhesion, Vol. 17, 1984, pp. 157-174.

84. Muller, Peter C., Stephen S. Kelley, and Wolfgang G. Glasser. "Engineering Plastics from Lignin. IX. Phenolic Resin Synthesis and Characterization", Journal of Adhesion, Vol. 17, 1984, pp. 185-206.

85. Myer, B. Urea-Formaldehyde Resins, Addison-Wesley Publishing Company, 1979.

86. Newman, William H., and Wolfgang G. Glasser. "Engineering Plastics from Lignin. XII. Synthesis and Performance of Lignin Adhesives with Isocyanate and Melamine", manuscript, Virginia Polytechnic Institute and State University, Department of Forest Products, and Polymer Materials and Interfaces Laboratory, 1985.

87. Newman, William H., and Wolfgang G. Glasser. "Engineering Plastics from Lignin. XII. Synthesis and Performance of Lignin Adhesives with Isocyanate and Melamine", manuscript, Virginia Polytechnic Institute and State University, Department of Forest Products, and Polymer Materials and Interfaces Laboratory, 1985.

88. Nguyen, Tinh, E. Zavarin, and E.M. Barrall II. "Quantitative Studies by Differential Scanning Calorimetry of the Reaction between Hydrogen Peroxide and Lignocellulose", Journal of Applied Polymer Science, Vol. 28, 1983, pp. 647-658.

89. Office of Technology Assessment. Wood Use: U.S. Competitiveness and Technology, GPO, 1983.

90. O'Laughlin, Jay and Ellefson, Paul V "U.S. Wood-Based Industry Structure-Part I-Top 40 companies," Forest Products Journal, Vol 31, No. 10, Oct. 1981, pp. 55-62.

91. Oliver, John F. Adhesion in Cellulose and Wood-based Composites, Plenum Press, 1981.

92. OPD Chemical Buyers Directory 1986, 73rd ed., Chemical Marketing Reporter, Schnell Publishing Company, 1985.

93. Pagel, Horst F., and Edwin R. Luckman. "EPI -- A New Structural Adhesive", Adhesives for Wood, edited by Robert Gellespie, Noyes Publications, 1984, pp. 139-149.

94. Panek, Julian R.; Cook, John Philip. Construction, Sealants and Adhesives, 2nd edition, John Wiley & Sons, 1984.

95. "Panel Industry Grows on World-wide Average" Word Wood, June 1984, pp. 21-25.

96. "Phenolics", Chemical and Engineering News, November 5, 1984, p. 11.

97. "Phenolic Resins -- Economy Is Key to Fortunes", Chemical & Engineering News, June 3, 1985, p.38.

98. "Phenolic Resins", Kirk-Othmer Concise Encyclopedia of Chemical Technology, Wiley-Interscience, 1985, pp.867-868.

99. "Phenolic Resin Producers Make Some Gains in Cost Pass-Along", Chemical Marketing Reporter, August 19, 1985, p.7+.

100. Philippou, John L., Eugene Zavarin, William E. Johns, and Tinh Nguyen. "Bonding of Particleboard Using Hydrogen Peroxide, Lignosulfonates, and Furfuryl Alcohol: Effects of Chemical Composition of Bonding Materials", Forest Products Journal, May 1982, pp. 55-62.

101. Philippou, John L., William E. Johns, and Tinh Nguyen. "Bonding Wood by Graft Polymerization. The Effect of Hydrogen Peroxide Concentration on the Bonding and Properties of Particleboard", Holzforschung, Vol. 36, No. 1, 1982, pp. 37-42.

102. Philippou, John L., William E. Johns, Eugene Zavarin, and Tinh Nguyen. "Bonding of Particleboard Using Hydrogen Peroxide, Lignosulfonates, and Furfuryl Alcohol: The Effect of Process Parameters", Forest Products Journal, March 1982, pp. 27-32.

103. Philippou, J.L., and E. Zavarin. "Differential Scanning Calorimetric and Infra-Red Spectroscopic Studies of Interactions between Lignocellulosic Materials, Hydrogen Peroxide, and Furfuryl Alcohol", Holzforschung, Vol. 38, No. 3, 1984, pp. 119-128.

104. Pizzi, A. ed. Wood Adhesives: Chemistry and Technology, Marcel Dekker, Inc., 1983.

105. Pizzi, A., C.J. Knauff, and P. Sorfa. "Effect of Acrylic Emulsions on Tannin-Based Adhesives for

Exterior Plywood", _Holz als Roh- und Werkstoff_, Vol. 39, 1981, pp. 223-226.

106. Plomley, K.F.   "Tannin-formaldehyde Adhesives for Wood, II. Wattle Tannin Adhesives", Division of Forest Products Technological Paper No. 39, Commonwealth Scientific and Industrial Research Organization, Australia, 1966.

107. "Production by the U.S. Chemical Industry", _Chemical and Engineering News_, June 10, 1985, p. 29.

108. Ramos, J.R., C.D. Olano, F.D. Chan, and M.P. San Pedro. "Kraft Black Liquor as Adhesive for Plywood", _NSDB Technical Journal_, April-June 1980, pp.42-49.

109. "Resorcinol Handbook", Koppers Company, 1982.

110. River, Bryan H., and Robert H. Gillespie. "Adhesives in Building Construction -- Past, Present, and Future", _Today's Problems -- Tomorrow's Opportunities_, The Adhesive and Sealant Council, 1980, pp. 31-42.

111. Robertson, Joe. "Extender Selection Becomes Critical as Glue Prices Rise" _Plywood and Panel Magazine_. March 1981, pp. 11-14.

112. Roy, D.C., P. Naha, and J. Nag.   "Tannin Adhesives for Plywood Manufacture -- Part II", _IPIRI Journal_, Vol. 4, No. 4, pp. 165-166.

113. Roy, D.C., P. Naha, and J. Nag.   "Tannin Adhesives for Plywood Manufacture -- Part III", _IPIRI_ Journal, Vol. 7, No. 1, pp. 1-4.

114. Roy, D.C., P. Naha, and B.N. Rajak.   "Cold-setting BWR Plywood Adhesive Based on Wheat Flour", _IPIRI Journal_, Vol. 5, No. 1, pp. 66-67.

115. Roy, D.C., A. Sarkar, and B.N. Rajak.   "Tannin Adhesives for Plywood Manufacture", _IPIRI Journal_, Vol. 3, No. 2, pp. 81-84.

116. Saayman, H.M., and J.A. Oatley. "Wood Adhesives from Wattle Bark Extract", _Forest Products Journal_, December 1979, pp. 27-33.

117. Schneberger, Gerald L. ed. Adhesives in Manufacturing, Marcel Dekker, Inc., 1983.

118. Selbo, M.L. Adhesive Bonding of Wood. Bulletin No.1512, U.S. Department of Agriculture, Forest Service, August 1975.

119. Sellers, Terry. Plywood & Adhesive Technology, Marcel Dekker, Inc. 1985.

120. Shah, R.S. "Utilisation of Black Liquor for Developing a Cheaper Adhesive for Plywood -- Part 1", Journal of the Society of Indian Forestors, January-June 1983, pp. 9-11.

121. Shen, K.C., and L. Calve. "Exterior Adhesive from Sulfite Mill Effluent", Adhesives for Wood, edited by Robert Gillespie, Noyes Publications, 1984, pp. 64-69.

122. Shen, K.C., L. Calve, and P. Lau. "A New Binder for Lignocellulosic Materials: Ammonium-Based Spent Sulfite Liquor", Proceedings of the Thirteenth Washington State University International Symposium on Particleboard, 1979, pp. 369-379.

123. Shields, J. Adhesives Handbook, 3rd edition, Butterworths, 1984.

124. Sittig, M. Practical Techniques for Saving Energy in the Chemical, Petroleum and Metals Industries, Noyes Data Corporation, 1977.

125. Skeist, Irving ed. Handbook of Adhesives, 1st and 2nd editions, Reinhold Publishing Corp., 1962, 1977.

126. Sorfa, P. "Some Practical Experience with Wattle Tannin Adhesives for Plywood in South Africa", Journal of Applied Polymer Science: Applied Polymer Symposium, 1984, pp. 63-67.

127. Steiner, P.R. "Adhesive and Extender Developments Using Foliage", Journal of Applied Polymer Science: Applied Polymer Symposium, 1984, pp. 145-157.

128. Steiner, Paul R., Suezone Chow, and Steven Vadja. "Interaction of Polyisocyanate Adhesive With Wood", Forest Products Journal, July 1980, pp. 21-24.

129. Stofko, John I. "Carbohydrate Transformation Bonding of Lignocellulosic Materials", Adhesives for Wood, edited by Robert Gillespie, Noyes Publications, 1984, pp. 54-63.

130. Swift, T. Kevin. Building Board Markets: Industry Study, Predicasts, Inc., May 1981.

131. UNIDO & FAO. First Consultation on the Wood and Wood Products Industry, UN Publications 1983.

132. U.S. Department of Agriculture, Forest Service. Adhesive Bonding of Wood, Technical Bulletin No. 1512, GPO, August 1975.

133. U.S. Department of Agriculture, Forest Service. An Analysis of the Timber Situation in the United States: 1952-2030, Forest Resource Report No. 23, December 1982.

134. U.S. Department of Agriculture, Forest Service, Forest Products Laboratory. "Selection and Properties of Woodworking Glues", Research Note FPL-0138 (Revised), October 1968.

135. U.S. Department of Agriculture, Forest Service. Structural Flakeboard From Forest Residues, General Technical Report WO-5, September 1978.

136. U.S. Department of Agriculture, Forest Service, Forest Products Laboratory. "Synthetic-Resin Glues", Research Note FPL-0141, June 1966.

137. U.S. Department of Agriculture, Forest Service. The Softwood Plywood Industry in the United States 1965 - 1982. Resource Bulletin FPL 13, Forest Products Laboratory, April 1984.

138. U.S. Department of Commerce, Bureau of Census. "Softwood Plywood", Current Industrial Reports, 1984.

139. U.S. Department of Commerce, Bureau of Census. "Fuels and Electric Energy Consumed: Part 1. Industry Groups and Industries", 1982 Census of Manufacturers, June 1983.

140. U.S. Department of Commerce, Bureau of Census. "Hardwood Plywood", Current Industrial Reports, 1984.

141. U.S. Department of Commerce. "Wood Products", U.S. Industrial Outlook 1985, GPO, pp.4-1 through 4-15, 1985.

142. U.S. International Trade Commission. Summary of Trade and Tariff Information: Synthetic Plastics Materials. USITC Publication 841, U.S. International Trade Commission, October 1977.

143. U.S. International Trade Commission. Summary of Trade and Tariff Information: Synthetic Plastics Materials. USITC Publication 841 (Second Supplement), U.S. International Trade Commission, December 1983.

144. Viswanathan, Tito. "Identification of Thermosetting Adhesive Resins from Whey Permeate as High Molecular Weight Maillard Polymers", Ind. Eng. Chem. Prod. Res. Dev., Vol. 24, 1985, pp. 176-177.

145. Wellons, J.D., and L. Gollob. "Molecular Properties of Phenolic Adhesives", Adhesives for Wood, edited by Robert Gillespie, Noyes Publications, 1984, pp. 26-31.

146. White, James T. "Growing Dependency of Wood Products on Adhesives and Other Chemicals", Forest Products Journal, November 1979, pp.14-20.

147. Wilson, James B. "Isocyanate Adhesives as Binders for Composition Board", Adhesives for Wood, edited by Robert Gillespie, Noyes Publications, 1984, pp. 134-138.

148. Wilson, James B. "Is There an Isocyanate in Your Future? -- Property and Cost Comparisons", International Symposium on Particleboard, Washington State University, 1981, pp. 185-193.

149. Woerner, Douglas L., and Joseph L. McCarthy. "Ultrafiltration of Kraft Black Liquor", Applications of Chemical Engineering in the Forest Products Industry, AIChE Annual Meeting, Los Angeles, CA, November 14-19, 1982, pp. 25-33.

150. Wu, King-Tsuen, and Eugene Zavarin. "Thermal Analysis of the Nitric Acid-Lignocellulose Mixtures", Journal of Applied Polymer Science: Applied Polymer Symposium, 1984, pp. 127-144.

151. Young, R.A., A. Krzysik, M. Fujita, S.S. Kelley, R.M. Rammon, B.H. River, and R.H. Gillespie. "Enhanced Wood Bond Strength Through Surface Treatment", summary paper to appear in Proceedings of the Adhesives '85 Symposium, U.S. Department of Agriculture, Forest Products Laboratory.

152. Young, Raymond A., M. Fujita, and B.H. River. "New Approaches to Wood Bonding: A Base-Activated Lignin Adhesive System", to appear in Wood Science and Technology, Springer-Verlag 1985.

153. Zoolagud, S.S., K.K. Mohandas, T.S. Rangaraju, T.R. Narayana Prasad, and Joseph George. "Tannin Extended Phenol-Formaldehyde Resin Adhesives for BWP Grade Plywood", IPIRI Journal, Vol. 5, No. 2, pp. 59-63.

Appendix C: Glossary

adhere, v :   to cause two surfaces to be held together by adhesion

adherend, n :   a body which is held to another body by an adhesive

adhesion, n :   the state in which two surfaces are held together by interfacial forces which may consist of valence forces or interlocking action, or both

adhesive, n :   a substance capable of holding materials together by surface attachment. Adhesive is a general term and includes among others cement, glue, mucilage, and paste. Various descriptive adjectives are applied to the term adhesive to indicate certain characteristics as follows:

         (1) Physical form - liquid, tape, etc.

         (2) Chemical type - silicate, resin, etc.

         (3) Materials bonded - paper, metal-plastic, can label, etc.

         (4) Conditions of use - hot-setting, etc.

assembly, n :   a group of materials or parts, including adhesive, which has been placed together for bonding or which has been bonded together

assembly time, n :   the ime interval between the spreading of the adhesive on the adherend and the application of pressure or heat, or both, to the assembly. For assemblies involving multiple layers or parts, the assembly time begins with the spreading of the adhesive on the first adherend. (see open assembly time and closed assembly time)

binder, n :   a component of an adhesive composition that is primarily responsible for the adhesive forces which hold two bodies together

black liquor, n :   in papermaking; liquid digester waste containing sulfonated lignin, rosin acids, and other wood waste components, from which tall oil is made

bond, n :   the union of materials by adhesives

bond, v :   to unite materials by means of an adhesive

bond strength, n :   the unit load applied in tension, compression, flexure, peel, impact, cleavage, or shear, required to break an adhesive assembly with failure occurring in or near the plane of the bond

caul, n: a sheet of material employed singly or in pairs in hot or cold pressing of assemblies being bonded. Cauls are employed usually to protect either the faces or the press platen or both against marring and staining, to prevent sticking, and to facilitate press loading. Cauls may be made of aluminum, stainless steel, hardboard, or fiberboard, with the length and width generally equal to the platen size of the press.

catalyst, n :   a substance that markedly speeds up the cure of an adhesive when added in minor quantity as compared to the amounts of the primary reactants

caul, n :   a sheet of material used singly or in pairs in hot or cold pressing of assemblies being bonded. Their purpose is to protect the assembly faces and the press platen against marring and staining, to prevent sticking, and to facilitate press loading. They can be made of aluminum, stainless steel, hardboard, or fiberboard, with the length and width equal to the platen size of the press.

closed assembly time, n :   the time interval between completion of assembly of the parts for bonding and the application of pressure or heat, or both, to the assembly (see open assembly time)

cohesion, n :   the state in which the particles of the adhesive (or the adherend) are held together by primary or secondary valence forces

condensation, n :   a chemical reaction in which two or more molecules combine with the separation of water or some other simple substance.

colloidal, adj n :   having the properties of colloid mater. Colloid matter has one or more of its dimensions in the range between one millimicron and one micron. In this size range, the surface area of the particle is so much greater than its volume that unusual phenomena occur. For example, the particles do not settle out of a suspension by gravity.

copolymer, n :   polymer where two or more monomers are involved

creep, n :   the dimensional change with time of a material under load, following the initial instantaneous elastic or rapid deformation

crosslinked, adj :   insoluble and infusible

cure, v :   to change the physical properties of an adhesive by chemical reaction, which may be condensation, polymerization, or vulcanication; usually accomplished by the action of heat and catalyst, alone or in combination, with or without pressure

cure time, n :   the period of time during which an assembly is subjected to heat or pressure, or both, to cure the adhesive

diluent, n :   an ingredient usually added to an adhesive to reduce the concentration of bonding materials

ether bond, n :   carbon-to-oxygen bond

extender, n :   a substance, generally having some adhesive action, added to adhesive to reduce the amount of the primary binder required per unit area

filler, n :   a relatively nonadhesive substance added to an adhesive to improve its working properties, permanence, strength, or other qualities

fiberboard, n :   a panel product manufactured from lignocellulosic fibers combined with a synthetic resin or other suitable binder. The panels are manufactured by the application of heat and pressure by a process in which the inter-fiber bond is substantially created by the added binder. Other materials may be added during manufacturing to improve certain properties.

flake, n :   a small wood particle of predetermined dimensions specifically produced as a primary function of specialized equipment. The cutting action is across the direction of the grain (either radially, tangentially, or at an angle) being such as to produce a particle of uniform thickness, essentially flat, and having the fiber direction essentially in the plane of the flakes. In overall character, flakes resemble a small piece of veneer.

flakeboard, n :   see particleboard

glue line (bond line), n :   the layer of adhesive which attaches to adherends

hardboard, n: a generic term for a panel manufactured primarily from interfelting lignocellulosic fibers (usually wood) consolidated under heat and pressure in a hot-press to a density of 31 pcf or greater, and to which other materials may have been added during manufacture to improve certain properties.

hardener, n :   a substance or mixture of substances to an adhesive to promote or control the curing reaction by taking part in it; also used to designate a substance added to control the degree of hardness of the cured film

hardwood plywood, n :   a plywood having at least three wood plies with a face of hardwood veneer; the back or inner plies may be any combination of hardwood or softwood veneer, lumber, particleboard, or hardboard.

hot press, n :   used in the preparation of particleboard, a press in which the platens are heated to a prescribed temperature by means of steam, electricity, or hot water

hydrophobic, adj :   incapable of dissolving in water. This property is characteristic of all oils, fats, waxes, and many resins.

hygroscopic, adj :	describes a substance that absorbs and loses moisture readily
inhibitor, n :	a substance that slows down chemical reaction. Inhibitors are sometimes used in certain types of adhesives to prolong storage or working life
laminate, n :	a product made by bonding together two or more layers of wood, either in straight or curved form, with the grain of all pieces essentially parallel to the length of the member
laminate, v :	to unite layers of material with adhesive
lamination, n :	the process of preparing a laminate; also, any layer in a laminate
lignin, n :	complex polymer of condensed phenylpropane. Lignins serve as the adhesive material of wood, cementing fibers and cells together to form the anatomical structure of wood. Lignins are susceptible to degradation by strong alkalis at high temperatures, by acid sulfite solutions at elevated temperatures, and by oxidizing agents. Lignins are removed from wood during the pulping process, leaving behind the cellulosic fibers used in papermaking.
lignosulfonate, n :	A metallic sulfonate (containing the $-SO_2OH$ group) salt made from the lignin of sulfite pulp-mill liquors.
methylol groups n. :	the molecular group $- CH_2OH$. Phenolic resole resins have this end group.
modifier, n :	any chemically inert ingredient added to an adhesive formulation that changes its properties
novolak, n :	a phenolic-aldehydic resin that, unless a source of methylene groups is added, remains permanently thermoplastic
open assembly time, n :	the time interval between the spreading of the adhesive on the adherend and the completion of assembly of the parts for bonding (see closed assembly time)
particleboard, n :	a generic term for a panel manufactured from lignocellulosic materials (usually wood) primarily in the form of discrete pieces or particles, as distinguished from fibers, combined with a synthetic resin or other suitable binder and bonded together by a process in which the entire inter-particle bond is created by the added binder, and to which other materials may have been added during manufacture to improve certain properties. Low density particleboards have a density of less than 37 pcf; medium density particleboards have a density between 37 and 50 pcf; high density particleboards have a density greater than 50 pcf. The majority of particleboard currently produced in the U.S. is medium density.
pcf:	particles per cubic foot
platen, n :	a steel plate constituting the pressure element in a hot press
plywood, n :	a glued wood panel made up of relatively thin layers, or plies, with the grain of adjacent layers at an angle, usually 90o. The outside plies are called facces or face and back plies; the inner plies are called cores or centers. The plies immediately below the face and back are called crossbands. The core may be veneer, lumber, or particleboard, with the total panel thickness typically not less than 1/16 inch or more than 3 inches. The plies may vary in number, thickness, species, and grade of wood.
polymer, n :	a compound formed by the reaction of simple molecules (monomers) having functional groups which permit their combination to proceed to high molecular weights under suitable conditions
post cure, n :	a treatment (normally involving heat) applied to an adhesive assembly following the initial cure, to modify specific properties
pot life :	See working life
reactive site, n :	an atom or group of atoms in a molecule that can form dipolar or covalent bonds with other atoms or molecules
reactivity, n :	indicator of the relative number of reactive sites and/or the relative potential bonding strength (degree of polarity) of the reactive sites. Can refer to a single molecule or a chemical substance.

resin, n :

a solid, semisolid, or pseudosolid organic material that has an indefinite and often high molecular weight, exhibits a tendency to flow when subjected to stress, usually has a softening or melting range, and usually fractures conchoidally

resol, n :

Phenol-formaldehyde resin formed by reacting phenol with excess formaldehyde (phenol: formaldehyde molar ratio greater than 1:1) in the presence of an alkaline catalysts. Does not require additional hardener (formaldehyde) for final curing to solid form.

solvent adhesive, n :

an adhesive having a volatile organic liquid as a vehicle (excludes water-based adhesives)

storage life, n :

the period of time during which a packaged adhesive can be stored under specified temperature conditions and remain suitable for use; also called shelf life

surface preparation, n :

a physical and/or chemical preparation of an adherend to render it suitable for adhesive joining

tannin, n :

a widely occurring group of substances that are of plant origin, derived largely from wood, nuts, bark, and tree leaves. It is a dark-colored polyhydroxy phenolic compound that varies in composition depending on its source. The most common commercial use for tannins is in the production of leather, where it combines with the proteins of raw hides to render leather. Tannin production involves the reduction of wood or bark into chips for soaking and boaling until the tannin is extracted.

technical lignin, n :

natural lignin modified through industrial processing

thermoplastic, adj :

capable of being repeatedly softened by heat and hardened by cooling

thermoset, adj. :

pertaining to the state of a resin in which it is relatively infusible

thermosetting, adj :

having the property of ujdergoing a chemical reaction by the action of heat, catalysts, ultraviolet light, etc., leading to a relatively infusible state; traditionally used for crosslinking compositions, even when they do not require elevated temperature to initiate chemical reaction

wood veneer, n :

a thin sheet of wood, generally within the thickness range from 0.01 to 0.25 inches (0.3-6.3 mm) to be used in a laminate

working life, n :

the period of time during which an adhesive, after mixing with catalyst, solvent, or other compounding ingredients, remains suitable for use

Acknowledgments

Hagler, Bailly & Company wishes to acknowledge the many members of the
adhesives and wood products community who contributed their knowledge,
experience, insights, and time to provide us with the information needed
to prepare this report. The following is a list of organizations contacted
during the course of this study.

DOMESTIC

UNIVERSITIES

Auburn University
Department of Forest
  Research
Forest Research Laboratory
Auburn, AL
Contacts: Mr. Leichti
          Dr. Bibliss
          Dr. Thomas Elder
          Dr. Ruen C. Tang

Colorado State University
Department of Forestry and
  Wood Science
Fort Collins, CO
Contact: Dr. Herb Schroeder

Michigan Technical University
Department of Forestry
Houghton, MI
Contact: Dr. P.E. Laks

Mississippi State University
Mississippi State, MS
Contacts: Terry Sellers
          Dr. Gary McGinnis

North Carolina State University
Department of Wood Science
Raleigh, NC
Contacts: Dr. Gratzl
           Dr. Chia Chen

Oregon State University
Department of Forestry
Corvallis, OR
Contact: Dr. Joseph Karchesy

Texas A&M University Department
  of Forestry
College Station, TX
Contact: Dr. Soltes

University of California
  Berkeley
Department of Forestry
Berkeley, CA
Contact: Dr. Eugene Zavarin,
              Professor of Forestry

University of Idaho
Department of Forestry
Moscow, ID
Contact: Dr. Alton Campbell,
              Assistant Professor

University of Illinois
Department of Forestry
Urbana, IL
Contact: Dr. Poo Chow,
              Professor

University of Wisconsin
Department of Forestry
Madison, WI
Contact: Dr. Raymond Young,
              Professor

Virginia Polytechnical
  Institute
Department of Forestry
Blacksburg, VA
Contact: Dr. Wolfang Glasser,
              Professor

Washington State University
Department of Wood Technology
Pullman, WA
Contact: Dr. Thomas Maloney,
              Professor

DOMESTIC

GOVERNMENT AGENCIES AND GOVERNMENT-SPONSORED LABORATORIES

Battelle, Pacific Northwest
  Laboratories
Richland, Washington
Contacts: David Johnson
          Janet Russell

Forest Products Laboratory
USDA Forest Service
Madison, WI
Contacts: Robert Gillespie,
           Project leader
           Al Christiansen

National Bureau of Standards
Building Materials Division
Gaithersburg, MD
Contact: Dr. Tinh Nguyen

National Science Foundation
Washington, DC
Contat: Dr. James Soo,
         Engineering Staff

Solar Energy Research Institute
Golden, CO
Contact: Dr. Helena L. Chum,
         Principal Scientist,
         Thermochemical and
         Electrochemical Research
         Branch

Southern Forest Products
  Experimental Station
Pineville, LA
Contacts: Dr. Richard Hemingway,
           Project Leader
           Dr. Chung Hse, Lignin
           Analyst

Tennessee Valley Authority
Knoxville, TN
Contact: Jenny Lee Cabler,
          Analyst - Biomass Branch

U.S. International Trade
  Commission
Washington, DC
Contact: Ed Taylor, Industry
          Analyst

TRADE ASSOCIATIONS AND PROFESSIONAL SOCIETIES

American Society for
  Testing & Materials (ASTM)
Philadelphia, PA
Contact: Tom Kochaba,
          Staff Manager for
          Committee D-14
          (Adhesives)

Forest Products Research
  Society
Madison, WI
Contact: Arthur B. Brauner,
          Executive Vice President

Hardwood Plywood Association
Reston, VA
Contact: Clark McDonald,
          President

Institute of Paper Chemistry
Appleton, WI
Contacts: Harry A. Posner, Jr.,
          President
          Dr. Werner Lonsky,
          Lignin Specialist

National Forest Products
  Association
Washington, DC
Contact: Jerry Prange,
          Director of Technical
          Services

National Particleboard
  Association
Gaithersburg, MD
Contact: Jerry Zinn,
          Technical Director

(formerly the Pulp Manufacturers
  Research League,which was
  merged into the Institute of
  Paper Chemistry, Division of
  Environmental Systems)
Appleton, WI
Contact: Averill J. Wiley,
          Consultant

The Society of Manufacturing
  Engineers, Adhesives Council
Dearborn, MI
Contact: Patricia Jones,
          Administrator

DOMESTIC

PRIVATE SECTOR

American Cyanamid Company
Polymer Products Division
Wayne, NJ
Contact: R.L. Vaughn,
           Marketing Manager

Ashland Chemical Company
Chemical Systems Division
Columbus, OH
Contact: Horst F. Pagel,
           Research Forest Products
           Technologist

Boise Cascade
Boise, ID
Contact: Dr. Alan Lambuth,
           Manager of R&D

Borden, Inc.
Borden Chemical Division
Contact: John Hine,
           Technical Director,
           Resins & Chemicals
           Division

Georgia-Pacific Corporation
Decatur, GA
Contacts: George Creech,
            Technical Specialist,
            Thermosetting
            Resin R&D
            Dr. J.D. Wellons,
            R&D Specialist

Hydrocarbon Research
Lawrenceville, NJ
Contact: Rick Johnson,
           Research Analyst

Koppers Company, Inc.
Chemicals and Allied Products Group
Pittsburgh, PA
Contact: M.I. Dietrick,
           Technical Director

Occidental Chemical Corporation
Durex Plastics Division
Niagra Falls, NY
Contact: Ray Knipple

Reed Lignin
Rothschild, WI
Contact: Steve Wylin,
           Research Manager

Reichhold Chemicals, Inc.
Specialty Phenolics Division
Warren, NJ
Contact: Bob Andrews,
           National Sales Manager

Skeist Laboratories
Livingston, NJ
Contact: Arnold Brink, Analyst

Union Carbide Corporation
Danbury, CT
Contact: Roy Schaufelberger,
           Sales Manager

Weyerhaeuser Technical Center
Tacoma, WA
Contact: Dr. Roland
           Kreibich, Consultant

INTERNATIONAL

AUSTRALIA

Division of Chemical & Wood
Technology
Commonwealth Scientific
  and Industrial Research
  Organization (CSIRO)
Highett Victoria, Australia
Contact: Dr. W.E. Hillis

AUSTRIA

Federal Forest Research Institute
Schonbrunn
Vienna, Austria
Contact: Hans Egger, Dipl. Ing.

BRAZIL

Brazilian Institute for Forestry
Development (IBDF)
Brazilia

Forestry Institute (Instituto
Florestal)
Sao Paulo, Brazil

Institute de Pesquisas Tecnologicas
do Estado de Sao Paulo
  (IPT)
Sao Paulo, Brazil

Instituto Nacional de Pesquisas
Agropecuarias (INPA)
Manuas-Am, Brazil
Contact: Dr. Harry Van Der Slooten

CANADA

Canadian Forestry Service
Ottawa, Ontario
Canada
Contact: Dr. R.J. Bourchier,
           Director General

Canadian Hardwood Plywood
Association
Canadian Particleboard Association
Ottawa, Ontario
Canada

Canadian Wood Council
Ottawa, Ontario
Canada

Council of Forest Industries of
British Columbia
Vancouver, British Columbia
Canada

Forintek Canada Corp.
Eastern Forest Products Laboratory
Ottawa, Ontario
Canada
Contacts: Dr. K.C. Shen
          Louis Calve,
          Research Scientist

Forintek Canada Corp.
Western Forest Products Laboratory
Vancouver, British Columbia
Contacts: Peter B. MacFarland,
           President
           P.R. Steiner,
           Adhesive Analyst

Iotech Corp.
Montreal, Canada

Lakehead University
Thunder Bay, Ontario
Canada

Laval University
Cite Universitaire, Quebec
Canada

National Research Council
Ottawa, Ontario
Canada

Reed Lignin
Quebec

Stake Technologies, Ltd.
Ottawa, Canada

Temfibre Inc.
Temiscamine, P.Q. Canada

University of British Columbia
Vancouver, British Columbia

University of Toronto
Toronto, Ontario
Canada

University of New Brunswick
Fredericton, New Brunswick
Canada

INTERNATIONAL

Xerox Research Center of Canada
Mississauga, Ontario
Canada

CHILE

Instituto Forestal
Santiago, Chile
Contact: Patricio Valenzuela
           Vasquez, Executive
           Director

Wood Technology Department
School of Agrarian, Veterinary and
   Forestry Science
University of Chile
Contact: Professor Emilio Cuevas
           Isquierdo

CHINA

Northeast Forest University
Harbin, PRC
Contact: Dr. Bao-Xue-Geng,
           Associate Professor
           of Wood Adhesives

FEDERAL REPUBLIC OF GERMANY

Bison-Werke Bahre & Greten
   GmbH & Co. KG
Federal Republic of Germany
Contact: Gunter Bucking,
           Leiter der
           Technologischen
           Entwicklung

Federal Research Center of Forestry
   and Forest Products
Institute of Wood Chemistry and
   Chemical Technology of Wood
Hamburg, Federal Republic of
Germany
Contact: Dr. J. Puls

Fraunhofer Institute of Wood
Research
Wilhelm Klaudits Institute (WKI)
Braunschweig, Federal Republic
   of Germany
Contacts: Dr. B. Dix
           Mr. R. Marutsky

Institute for Wood Chemistry
Hamburg University
Hamburg, Federal Republic of
Germany
Contact: Dr. G. Weissmann

Bundesforschungsantalt fur Forst-
   und Holzwirtshaft
Federal Republic of Germany
Contact: Dr. Cihan Ayla

Polymer Institute
University of Karlsruhe
Federal Republic of Germany
Contact: Mr. Horst H. Nimz

FINLAND

Finnish Forest Research Institute
(Metsantutkimuslaitos)
Helsinki, Finland
Contact: Dr. Olavi Huikari,
           Director-General

Finnish Pulp and Paper
   Research Institute
(Oy Keskuslaboratorio --
   Centrallaboratorium Ab)
Helsinki, Finland
Contact: Dr. Kaj Forss

Technical Research Center of
Finland
Forest Products Laboratory
Espoo, Finland
Contact: O. Liiri

University of Oulu
Department of Chemistry
Oulu, Finland
Contact: Dr. Erkki Pulkkinen

FRANCE

CdF Chimie
Paris, France
Contact: Mr. Lechattre

Centre Technique du Bois (CTB -
   Center of Wood Technology)
Paris, France
Contacts: Mr. Gaillard
           Mr. Elbez

INTERNATIONAL

Franet
Senlis Cedex B.P. 100
France
Contact: Denis Morat,
   Marketing Manager

Institut National des Recherches
   Agronomiques
(INRA - National Institute of
   Agricultural Research)
Paris, France
Contact: Mr. Montis

ISOROI
Lisieux, France
Contact: Mr. Barray,
           Director of
           Technical Services

Laboratoire Rousselot
Choisy le Roi, France
Contact: Mr. Carriere

Rougier, Ocean, Landex (ROL)
NIORT, France
Contact: Mr. Jean-Claude Coue,
           Research Director

Service de Biochimie Appliquee
(Applied Biochemistry
   Department)
Ecole Nationale Superieure des
   Etudes
Agronomiques et Industries
   Alimentaires
(ENSAIA - National School for
   Advanced Studies of
   Agriculture and Food Industries)
Vandoeuvre, France
Contact: Mr. Metche

INDIA

Composite Wood Branch
Forest Research Institute
Dehra Dun, India
Contact: Mr. R.C. Gupta

Indian Plywood Industries
   Research Institute (IPIRI)
Bangalore, India
Contact: Dr. P.M. Ganapathy,
           Director

INDONESIA

Forest Research Institute
Jalan Gunung Batu
Bogor, Indonesia
Contact: Mr. Soerjono

IRELAND

Loctite International Research &
   Development Laboratory
Dublin, Ireland
Contact: Dr. R.S. Charnock

ITALY

FAO
Rome, Italy
Contact: F.J. Keenan, Director,
           FOI

Wood Institute
Florence, Italy
Contact: Mr. Attilio Arrighetti,
           Director

JAPAN

Department of Agriculture
Tokyo University of Agriculture
   and Technology
Tokyo, Japan

Forestry and Forest Products
   Research Institute
Ibaraki, Japan
Contact: Dr. Mitsuma Matsui,
           Director General

University of Tokyo
Department of Agriculture
Tokyo, Japan
Contact: Dr. Bunichiro Tomita,
           Faculty of Agriculture

Wood Research Institute
Kyoto University
Japan
Contact: Dr. Takayosohi
           Higuchi, Director

Hokkaido Forest Products
   Research Institute
Hokkaido, Japan
Contact: Mr. Sumio Imura

INTERNATIONAL
_____

NETHERLANDS

Forest Products Research Institute
TNO
Delft, The Netherlands
Contact: Ing A. van der Velden,
          Director

NEW ZEALAND

CSIR
Chemistry Division
Petrone, New Zealand
Contacts: L.Y. Foos
          Dr. L.J. Porter

Forest Research Institute
Wood Technology Division
Rotorua, New Zealand
Contacts: Dr. C. Bassett,
           Director of Research
           David Plackett

New Zealand Forest Products Limited
Auckland, New Zealand
Contact: Dr. Wood

NIGERIA

Ahmadu Bello University
Department of Chemical Engineering
Zaria, Nigeria
Contact: R.O. Ebewele

NORWAY

Dyno Industrier A.S.
Chemicals Group
Oslo, Norway
Contacts: Ken P. Harris
          Per Hanetho, Manager,
          Adhesive Resins
          Laboratory

PHILIPPINES

Forest Products Research
  & Development Institute
National Science and Technology
  Authority
College, Laguna
Republic of the Philippines
Contacts: Felisa D. Chan,
             Sc. Res. Specialist
           Mrs. Adelina E. Manas,
             Sr. Sc. Res. Specialist
           Mrs. Erlinda C. Salud,
             Sr. Sc. Res. Specialist
           Mrs. Mildred Fidel,
             Sc. Res. Specialist

SOUTH AFRICA

Bondtite Adhesives
Republic of South Africa
Contacts: Mr. A Ryder,
             Technical Director
           Dr. E.J. Sheppard,
             Wattle Industries Centre

General Chemical Corporation
Johannesburg, Republic of South
Africa
Contact: Mr. Terence Martin
          Goulding

National Timber Research Institute
Pretoria, Republic of South Africa
Contacts: Dr. D.L. Bosman, Director
          Dr. A. Pizzi, Chief
          Director

Protea Chemical Manufacturing
Johannesburg, Republic of South
Africa
Contact: T.M. (Martin) Goulding,
          Technical Manager

The University of the Orange
  Free State
C.S.I.R. Flavonoid Chemistry
  Research Unit
Department of Chemistry
Bloemfontein, Republic of South
Africa
Contact: D.G. Roux, Professor
          of Organic Chemistry

INTERNATIONAL

SOVIET UNION

Leningrad Academy of
  Wood Technology
Contact: A.A. Elbert

Tallin Polytechnic Institute
Contact: Dr. Kissler

SWEDEN

Swedish Institue for Wood
  Technology Research
(Institutet for Trateknisk
  Forskning)
Stockholm, Sweden
Contact: Ingvar Johansson

TAIWAN

Forest Products Division
Department of Forestry
National Chung Hsing University
Taichung, Taiwan

Taiwan Forestry Research Institute
Taipei, Taiwan
Contact: Shen-Chen Liu, Director

UNITED KINGDOM

The British Adhesives and
  Sealants Association
Hythe, Southampton
Hants, England
Contact: Dr. Peter Bosworth

The City University
Adhesion Science Group
London, England
Contact: Prof. Allen

Imperial Chemical Ind. Ltd. (ICI)
Blackley, Manchester
England
Contact: G.W. Ball, Deputy
          Section Manager

Princes Risborough Laboratory
Department of the Environment
Building Research Establishment
Aylesbury, Bucks
Contacts: John Brazier
           Bryan Paxton
           John M. Dinwoodie

Timber Research and
  Development Association
Hughenden Valley, High Wycombe
Bucks
Contact: Dr. Gavan Hall

University of North Wales
Department of Wood Science
Contact: Dr. Banks